Everyday Math for the Numerically Challenged

By
Audrey Carlan

CAREER PRESS
3 Tice Road
P.O. Box 687
Franklin Lakes, NJ 07417
1-800-CAREER-1
201-848-0310 (NJ and outside U.S.)
Fax: 201-848-1727

℘

EVERYDAY MATH FOR THE NUMERICALLY CHALLENGED
Cover design by Foster & Foster
Printed in the U.S.A. by Book-mart Press

To order this title, please call toll-free 1-800-CAREER-1 (in NJ and Canada: 201-848-0310) to order using VISA or MasterCard, or for further information on books from Career Press.

Library of Congress Cataloging-in-Publication Data

Carlan, Audrey.
    Everyday math for the numerically challenged / by Audrey Carlan.
       p.   cm.
    Includes index.
    ISBN 1-56414-355-4 (pbk.)
    1. Mathematics--Popular works.   I. Title.
  QA93-C37   1998
  510--dc21   98-10725

Contents

Part 2: Charts and Graphs

Part 3: Using Algebra

Introduction

Hi there! Let's get to know one another. That's what an introduction is for.

How do you do, dear reader. I'm Audrey Lee, a Professor Emeritus (that means retired with merit) of Math and Computer Science in the Community Colleges.

I taught math in the California community colleges for more than 23 years. Among the courses I taught were Arithmetic for College Students, PreAlgebra, and Elementary Algebra. I presented my students with the basic concepts, always in a simple, step-by-step manner.

I observed that students can understand basic concepts of arithmetic and algebra if these concepts are developed slowly and *all* the details are filled in (not skipping any "obvious" steps). We used simple words to explain the math concepts; new math words were always carefully defined. I have placed such new math words in the glossary in Appendix I on page 125.

Thus, the students moved along from concept to concept and did not get "lost." *In fact, they started to actually enjoy the logic of arithmetic!*

Now that I have retired, I observe that many adults can benefit from a good, *easy-to-read*, step-by-step explanation of how to understand and solve everyday math problems.

I have divided *Everyday Math for the Numerically Challenged* into three parts:

Part 1: Using Percent
Part 2: Using Charts and Graphs
Part 3: Using Algebra

Part 1 introduces everyday percent problems, solved for you in a step-by-step manner. There are nine different types of percent problems that you will probably find relevant to your needs.

Part 2 introduces you to pie charts, bar graphs, and line graphs and presents an assortment of everyday problems that can be illustrated using these graphing tools.

In Part 3, you will be introduced to some of the language of Algebra: constants, variables, and algebraic expressions. Also, you will work with everyday formulas, and some logic that is inherent to algebra. For numerically challenged readers, I give a short explanation of probability and how it is applied to the game of poker!

In all three parts, I encourage the use of hand calculators. I will show you how to use the hand calculator to solve percent problems.

This book has been written for *you*, Dear Reader, not for any math majors or math professors. Practical problems will be presented throughout the book, followed by solutions.

Why do we need all these skills? The more complex our world becomes, the more important it is for all of us to understand how to do some of the necessary problem-solving that arises in our daily lives. We encounter interest, loans, sales tax, income taxes, commissions, sale mark-downs (discounts on cars, clothing, food, etc.), all of which require the use of percentage. Lists of various ingredients in packaged foods, vitamins, food supplements, etc., are given in percent; housing costs (selling, buying), professional fees (lawyers, real estate brokers), percent profit and losses, percent grades in classes, and many other day-to-day matters including trying to understand the income tax forms.

We are constantly being shown facts and figures, charts and graphs. Lots of times, our eyes glaze over, and we just nod our heads and let someone else figure out the solutions.

Sales taxes on many store items are already calculated by the computers at the check-out. But what should happen if, for example, the electric power went out, and all the check-out computers "went down" (computerese for would not work)? Do we buy the merchandise? Do we pay the sales taxes? If we need a lawyer, or a real estate professional, do we know how to calculate the fees? Are we getting the correct discount on the used car lot? Are we paying the correct interest on our loan?

It's time for us to understand the basics, so that we may be able to better understand these solutions to everyday math problems.

What do you have to know before getting started? You should know how to add, subtract, multiply, and divide whole numbers (or know how to do those arithmetic operations on a hand calculator).

What else do you need? A positive outlook, and a little patience. *Together, we can do it!*

Part 1: Using Percent

Chapter 1
An Introduction to Fractions

Why should we be concerned with fractions? Because they are *there*! No fooling. Just as long as industry, consumers, and government still use certain fractional weights and measures (called the "British System of Measure"), and *not* Metric System (which is a decimal system), then, my dear reader, we should be aware of and have some degree of knowledge about fractions.

Do you remember how to add, subtract, multiply, and divide fractions? (Was that an "Ugh" I heard?) Well, there is help in this electronic age. Ordinary hand calculators don't calculate fractions; you have to be able to change fractions into decimals (we will learn that method in Chapter 4) in order to use a hand calculator to calculate fractions. However, I do know of several brands of hand calculators that have the capability to add, subtract, multiply, and divide fractions. They have a special key that looks like:

a b/c

I have been using such calculators for years. It has made working with fractions an easy task.

Before we go any further, let's review what a fraction is: *A fraction is the ratio or comparison of two numbers*, written as A/B (read this as "A over B") or $\frac{A}{B}$ where the value of B is *never* zero.

The value of A is called the *numerator*, and the value of B is the *denominator*.

The fraction A/B can also be read as "A divided by B," or A ÷ B. Think of the fraction this way: You have a certain object (say, a pizza pie). You cut it into 8 *equal* pieces. Therefore, the *denominator* is 8. Now, you eat 4 pieces of the pie. Thus, the *numerator* is 4. You have eaten 4/8 of the pizza. (That is, four equal pieces out of the whole pizza pie of eight equal pieces.)

You can eat one piece: 1/8 of the pizza; or two pieces: 2/8 of the pizza; three pieces: 3/8 of the pizza; four pieces: 4/8; and so on. When you eat the whole pie, all eight pieces, you will have eaten 8/8 parts of the pizza. The whole pizza is 8/8 pieces thus 8/8 = 1.

If you were to cut the pizza pie into 6 pieces, you could eat one piece: 1/6; two pieces: 2/6; three pieces: 3/6; and so on. The whole pie is all six pieces: 6/6. The whole pizza is 6/6 pieces, thus 6/6 = 1.

Notice that when the fraction has the same numerator and denominator we can write it as A/A, where A can be any number other than zero. The fraction A/A represents the whole number 1 (the whole pie). Thus: B/B = 1; 2/2 = 1; 17/17 = 1, and so on.

Here are some fancy words about fractions to impress your friends with.

Proper fractions: fractions that have numerators *smaller* than their denominator. For example: 1/2, 3/4, 5/8.

Improper fractions: fractions that have numerators *larger or equal* to their denominator. For example: 7/4, 9/5, 6/6.

Equivalent fractions: fractions such as 3/6 and 4/8 that represent *the same amount*. That is, if you had eaten 3 pieces out of a pizza pie that had been cut into 6 pieces, you would have eaten the *same amount* as if you had cut the same pie into 8 pieces and had eaten 4 of the pieces! So, 4/8 and 3/6 are called *equivalent fractions*.

An important skill we need to know when working with fractions is how to reduce fractions to lowest terms. First, let us understand what "lowest terms" means. Let us say you can divide the numerator and the denominator of a fraction evenly by a common non-zero factor. (Factors are numbers that multiply each other.) Then, when you cannot divide the numerator and denominator any further, you have *reduced the fraction to lowest terms*.

In other words, you look for common "factors" in the numerator and denominator. You remember from our pizza pies that 6/6 = 1 and 8/8 =1. Now you will see how to reduce a fraction to lowest terms.

Consider the fraction 25/100. The numerator and denominator of this fraction have factors in common. You can factor the numerator, and separately factor the denominator. Remember, when you have the fraction A/A where A is not zero, the value of the fraction is 1. You group the factors as shown. The fraction reduces to 1/4.

$$\frac{25}{100} = \frac{5 \cdot 5}{5 \cdot 5 \cdot 2 \cdot 2} = \frac{5}{5} \cdot \frac{5}{5} \cdot \frac{1}{2 \cdot 2} = 1 \cdot 1 \cdot \frac{1}{2 \cdot 2} = \frac{1}{4}$$

Consider the fraction 75/100. Again, the numerator and denominator have a factor in common.

You can multiply *or* divide both the numerator and denominator by the same nonzero number to reduce to lowest terms. In this case, you can divide both the numerator and denominator by 25:

$$\frac{75}{100} = \frac{75 \div 25}{100 \div 25} = \frac{3}{4}$$

When you reduce fractions to lowest terms, try to find numbers that go into both numerator and denominator evenly. Reduce 125/100 to lowest terms. Notice that this fraction is an improper fraction because the denominator is less than the numerator. These improper fractions can also be written as mixed numbers.

$$\frac{125}{100} = \frac{125 \div 25}{100 \div 25} = \frac{5}{4}$$

Mathematically, the fraction 5/4 can be written as:

$$\frac{5}{4} = \frac{4+1}{4} = \frac{4}{4} + \frac{1}{4} = 1\frac{1}{4}$$

As a shortcut, the easiest way to write 5/4 as a mixed number is to divide the denominator into the numerator. When you do this, the numerator is called the dividend, and the denominator is called the divisor (because it divides into the numerator). The answer is called the quotient.

To change the improper fraction 5/4 to a mixed number write the expression as:

$$\frac{5}{4} = 4\overline{\smash{)}5} = 1\frac{1}{4}$$

In the next chapter, you will see how the fractions are another way of expressing decimals.

Steve orders a small pizza in a restaurant.

The waiter asks: "Do you want your pizza cut into 4 or 6 pieces?" Steve thinks for a minute, and then replies: "Better make it 4 pieces; I don't think I can eat 6!"

Chapter 2
A Brief Discussion of Decimals

We are all concerned with decimals, especially when we see numbers with $ signs, such as $52.89, $125.09, and the like. All the operations we do on ordinary hand calculators are in decimal numbers. Later, we will see a direct relationship between fractions and decimals, as well as percents. Once we get these relationships organized, you will find them quite simple to understand. It's almost like understanding your family tree and how you are related to some of your relatives.

Now we will learn about decimals, and the relationship between fractions and decimals. In Chapters 3 and 4, we will change fractions to decimals and decimals to fractions.

You will see how math logic flows, from one basic concept to the next. Relax, and flow with the math!

Here are two basic rules you should remember about decimals:

1. *A decimal number is a fraction whose denominator is a power of ten.*
2. *All whole numbers are decimal numbers with an understood decimal point to the extreme right of the number.*

What are "whole numbers"? They are natural counting numbers (you know, 1,2,3, etc.) and 0.

Whole numbers can be considered decimal numbers that have an understood decimal point to the extreme right of the number. (That means the decimal point is there, we do not see it or have to write it.) For example 15 is a whole number. The decimal 15.0 represents the same number as the whole number.

What is meant by "power of 10"? Here, we will introduce some shortcut notations to make it easier (honest!) to understand the decimal system.

Multiply $10 \cdot 10 = 100$; $10 \cdot 10 \cdot 10 = 1,000$; $10 \cdot 10 \cdot 10 \cdot 10 = 10,000$; and $10 \cdot 10 \cdot 10 \cdot 10 \cdot 10 = 100,000$, so that, we have:

$10 \cdot 10$ contains *two equal factors*. (Factors are numbers that multiply each other.)
$10 \cdot 10 \cdot 10$ contains *three equal factors*.
$10 \cdot 10 \cdot 10 \cdot 10$ contains *four equal factors*.
$10 \cdot 10 \cdot 10 \cdot 10 \cdot 10$ contains *five equal factors*.

Someone thought it would be neat (and precise) to use a shorthand notation to show these factors as follows:

10: for one factor, show the one factor as 10^1 (in this case, the "1" is always understood, so that we can just write and think 10).
$10 \cdot 10$: for two equal factors show as 10^2.
$10 \cdot 10 \cdot 10$: for three equal factors show as 10^3.

We say that the *number 10* is called the *base* and the number that tells how many equal factors there are is called the *exponent* or *power*. The exponent is always written above and to the right of the base (a superscript).

Therefore, we can write some powers of ten as: 10^2, 10^3, 10^4, etc.

What would 10^6 represent?

$10^6 = 10 \cdot 10 \cdot 10 \cdot 10 \cdot 10 \cdot 10 = 1,000,000.$

Did you notice that when you multiply any number by 10, you simply add one zero to (the right side of) the whole original number. This means that when you multiply any number by a power of 10, you add the number of zeros as represented by the exponent to the right side of the original whole number.

Thus: $4 \cdot 10^4 = 40,000$, and $6 \cdot 10^6 = 6,000,000.$

Now, let's get back to decimal numbers. *A decimal number is a fraction whose denominator is a power of ten.* For example, the decimal number 5.1 is a fraction that can be written as 5 1/10. In each case, the number is read as "five and one tenth."

Our decimal number system is set up as shown in the Decimal System Place Values table on page 9. The table is laid out with the columns labeled: Hundred, Ten, Unit, Decimal Point, Tenth, Hundredth. When we place *digits* into the correct column of this table, we get our *decimal* numbers.

Our decimal number system is called "decimal" from *dec*, meaning 10. As shown in the following table, the columns are powers of 10. Notice that the Units place is represented by 10^0. This definition makes for a nice consistent decimal system place value.

Digits are any one of the Arabic numeral symbols: 0,1,2,3,4,5,6,7,8,9.

Consider the following decimal numbers, and place them in the following Decimal System Place Value chart: 123.45, 36.72, 195.06, 202.10, and 17.00. Line up the digits on the decimal point column.

Decimal System Place Values

Hundred 10^2	Ten 10^1	Unit	Dec Pt	Tenth $1/10^1$	Hundredth $1/10^2$
1	2	3	.	4	5
	3	6	.	7	2
1	9	5	.	0	6
2	0	2	.	1	0
	1	7	.	0	0

Let us set up fractions whose denominators are powers of 10: 7/10; 12/100; 45/1000. These are proper fractions, and are read as follows:

7/10 reads as seven over ten or *seven tenths.*

12/100 reads as twelve over one hundred or *twelve* hundredths.

45/1000 reads as forty five over one thousand or *forty five thousandths.*

Now, here is the connection between fractions and decimals. Decimals are fractions whose denominators are powers of 10.

When we say 7/10 (seven tenths), we write the number as *0.7* (one decimal place). When we say 12/100 (twelve hundredths), we write the number as *0.12* (two decimal places). When we say 45/1000 (forty five thousandths), we write the number as *0.045* (three decimal places). When we say 125/1000 (one hundred twenty five thousandths), we write the number as *0.125* (three decimal places).

Do you see? The *exponent* of the *base 10* in the denominators now tells us how many decimal places we use after the decimal point!

Now you can understand that $9/1,000,000 = 9/10^6$ $= 0.000009$ because there are six decimal places in the number. This number is read as: 9 over one million, or 9 millionths.

We know that our everyday math problems have fractional or decimal numbers. Some very important ones have both types of numbers.

How can we compare prices of similar items (comparison shopping) in a supermarket when items come in various sizes (pounds, ounces, *fractional* pounds, *fractional* ounces), and the price is a *decimal* number? Or how about the prices of yard goods (cloth, drapery materials, carpeting) when they are sold by the yard and the price is a decimal number? In Appendix 2, you will understand how to become a better shopper by determining unit pricing as well as the prices for yard goods. In addition, you will have a practical example, working with decimals (dollars and cents) to balance and reconcile your checking account.

Chapter 3
Changing Decimals to Fractions

Fractions can be expressed as decimals, and decimals can be expressed as fractions. They each represent the same number in a different form. Here we will change decimal numbers to fractions.

If you are given a decimal, first read the number (either aloud or silently).

For example: 0.235 is read: two hundred thirty five *thousandths*. Thus, to write this as a fraction, the number 235 now becomes the *numerator* and the thousandths becomes the *denominator*.

$$0.235 = 235/10^3 = 235/1000$$

The decimal number 0.0255 is read: two hundred fifty five *ten thousandths*. As a fraction, the number 255 now becomes the numerator, and the ten thousandths becomes the denominator: $0.0255 = 255/10^4 = 255/10,000$.

The number 16.25 is read: sixteen and twenty five hundredths. It is written as 16 25/100. This number contains the whole number 16 and the fractional part 25/100. This type of number, composed of a whole number and a fractional part, is called a *mixed number*. (Note that mixed numbers can also be worked on hand calculators that do fractional arithmetic.)

Some readers may remember that they were told to always reduce fractions to lowest terms. The above examples only illustrate the *relationship* between decimals and fractions. We discussed reducing fractions to lowest terms in Chapter 1.

Once you have changed from a decimal to a fraction, and understand the process and the relationship between the fraction and decimal, then you can reduce the fraction to lowest terms as shown in Chapter 1.

Consider the decimal number 12.45. This can be written as a whole number and a fractional part, 12 and 45/100. Thus we can write:

$$12.45 = 12 \frac{45}{100}$$

In Chapter 1 we discussed ways to reduce fractions to lowest terms. Perhaps the easiest way is:

$$\frac{45}{100} = \frac{5 \cdot 9}{5 \cdot 20} = \frac{5}{5} \cdot \frac{9}{20} = \frac{9}{20}$$

Thus, 12 45/100 = 12 9/20.

Chapter 4
Changing Fractions to Decimals

Changing fractions to decimals will require a little arithmetic calculation. You can do this easily on a hand calculator as follows: Key in the numerator, press the ÷ button, key in the denominator, press =. The decimal number will appear in your display.

For example, you can change the fraction 1/8 to a decimal using a hand calculator. Key in 1, press ÷ symbol, key in 8, press the = symbol. The display should read: 0.125. Any fraction A/B (where B is not zero) can be changed to a decimal number in this manner.

(As you may know, you cannot divide by zero. Try this: Key in 1 then press ÷ key, then key in 0, and press =. We went through the steps of this division, but your display may be reading "Error," or even lock itself so that you have to clear the calculations! Calculators know there is no numerical answer to division by zero!)

You can do the same calculation by hand! You divide the Denominator *into* the Numerator. For example, if the fraction is 3/4, you divide 4 into 3 so that:

$$4\overline{)3.00} = 0.75$$

Thus, the fraction 3/4 is equal to the decimal number 0.75.

HERE IS A RIDDLE:

A rancher died. He left his 17 horses to his 2 sons and 1 daughter. His last request was that his daughter, Joy, was to have 1/2 of the horses; his son Jim was to have 1/3 of the horses, and his son Jon was to have 1/9 of the horses. The children were confused and could not figure out how to divide the horses among them without getting fractions of the horses!

They called their rancher neighbor, Old Sam, who said he would come right over to help them.

Old Sam came by, riding his horse Old Paint, and he let Old Paint out into the pasture with the other horses. Now there were 18 horses in the pasture!

"What's the problem?" he asked. Old Sam gave 1/2 of the horses in the pasture, that was 9 of them, to Joy. He gave 1/3 of the horses, that was 6 of them, to Jim, and he gave 1/9 of the horses, 2 of them, to Jon. Then, Old Sam jumped on Old paint, and rode home.

(The original fractions set up by the rancher add up to 17/18. There were 17 horses. When Old Sam came by with his horse, and put Old Paint in the pasture, there were 18 horses! Now, the fractions could be worked with whole numbers of horses to be given, and Old Paint, the 18th horse was ridden back home again by Sam.)

Chapter 5
Truncating and Rounding Decimals

You may hear on the radio or on TV, or read about some amounts as "approximately $12,000" or that "the crowd was estimated at approximately 10,000." You say you have about $15 in your pocket (even though you really have $15.47).

In many applications using decimal numbers, you may want to approximate or you may only need a few decimal places (the places to the right of the decimal point). You would then have to either *truncate* or *round* the decimal number to conform to the requirements. Rounding and truncating are two techniques that allow for satisfactory accuracy, yet shorten the resulting decimal places in the final answer.

Hand calculators (even though they may display 8 or more decimal places) truncate the numbers if there are more decimal places in the answer than can fit on the display.

To *truncate* a number means to *drop off or remove* all digits past a certain specified number place.

For example, to truncate 25.45798 to *three* decimal places, remove all digits past the third decimal place. 25.45798 truncates to 25.457. Mathematically, we write this as 25.45798 ~ 25.457, where the ~ symbol means "approximately equal to."

Here are a couple more examples.

Truncate 15.45798 to a whole number. 15.45798 truncates to 15; write this as: 15.45798 ~ 15, a whole number.

Truncate 295.4062 to tenths. 295.4062 truncates to 295.4; write this as 295.4062 ~ 295.4, our number truncated to tenths.

Rounding provides a better degree of accuracy than truncating. Here is how to round decimal numbers:

3. Locate the rounding place desired.
4. Look at the digit *to the right of* the rounding place.
5. If that digit is 5 or more, increase the rounding place number by 1; if that digit is less than 5, *do not* change the rounding place number.

After you have done steps 1 through 3:

6. Locate the decimal point in the number.
7. If the rounding place number is *to the right* of the decimal point, drop all digits to the right of the rounding place number.
8. If the rounding place number is *to the left* of the decimal point, *replace all digits to the right of the rounding number, and up to the decimal point,* with zeros. (This will give you whole numbers.)

This may sound complicated, but a few examples will show you that this method is easy to understand.

Round 6.1834 to the nearest tenth.

Locate the tenths place: 6.1834. The digit to the right of the underlined number is more than 5. Therefore, increase the tenths place number (1) by 1 since the rounding place number is to the right of the decimal point (step 5). 6.1834 rounds to 6.2, or we write: 6.1834 ~ 6.2.

Round 257.721 to the nearest whole number (units place).

Locate the units place: 25_7.721. The digit to the right of the underlined number is more than 5. Increase the rounding place number (7, the units place) by 1. The rounding place number is to the left of the decimal point (step 6). 25_7.721 rounds to 258, we write 25_7.721~ 258.

Round 23,945 to the nearest thousand place.

Locate the thousand place: 2_3,945. The digit to the right of the thousand place digit is greater than 5. Increase the rounding place digit so that it is a 4. Now follow steps 4,5, and 6. Replace *all* digits to the right of the rounding place digit with zeros, so that: 2_3945 rounds to 24,000, so that we write: 23,945 ~ 24,000.

Here are a few more examples. Round the following numbers, as indicated:

Round 56.8662 to the nearest tenth.
56.8662 ~ 56.9
Round 7.3999 to nearest whole number.
7.3999 ~ 7
Round 9,479.3 to the nearest hundred.
9,479.3 ~ 9500
Round 7.008 to the nearest hundredth.
7.008 ~ 7.01

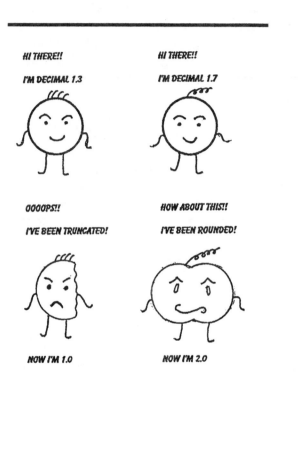

Chapter 6
Linking Decimals to Percents

Take another look at your hand calculator. Your calculator may have a % operator key. However, that key operator does not change numbers from decimal to percent or percent to decimal. The % operator, however, will allow you to work with percents, as I will explain later.

However, we should know the connection between decimals and percents and how to change from one to the other.

Percent means *Parts Per Hundred*, and is represented by the symbol %. Percent means hundredths, and hundredths as a decimal is written using *two* decimal places. We can use the symbol % in place of two decimal numerals in the number.

When we have exactly two decimal places in the number we move the decimal point two places to the right, and put in the % sign, so that 0.25 = 25%.

Notice that for decimals that have values less than 1 (written as: < 1), we put a zero in front of the decimal point, as shown with 0.25 above.

You can see that there is a percent symbol (per hundred), and *no* decimal places in 25%. The decimal point is replaced by the % symbol.

Some other examples are: 0.68 = 68%;   0.06 = 6%; 0.55 = 55%;   0.10 = 10%.

When we have more than two decimal places in the number we still move the decimal point the two places, but we must remember to insert a decimal point where it is needed.   0.575 = 57.5%; 0.1625 = 16.25%; 0.005 = 0.5%; 0.0005=0.05%.

The percent symbol accounts for two of the decimal places. We still must account for the other decimal places in our answer. Thus we have: 0.068 = 6.8%; 0.3555 = 35.55%; 0.10005 = 10.005%.

Here is an easy rule to remember when changing from decimal to percent: *Move the decimal point two places to the right, and put in the % symbol.*

Let us consider whole numbers, and how to change them to percents. First we must realize that whole numbers have an *understood* decimal point to the *right* of their last digit. The number 12, for example, can also be written as 12.0, or 12.00000, with as many zeros after the decimal point as we wish. Thus, 12 = 12.000 and 345 = 345.000 and 72 = 72.000.

Now we can change numbers with decimals to percent, using the method explained above. The rule is: *Move the decimal point two places to the right, and put in the % symbol.* For example:

$$12 = 12.000 = 1200.0\%$$
$$345 = 345.000 = 34500.0\%$$
$$72 = 72.000 = 7200.0\%$$

In these above percents, the zero to the right of the decimal point may be omitted. That is:

12 = 1200%; 345 = 34500%; and 72 = 7200%.

Changing percents to decimals is quite easy. If you have followed the method for changing a decimal to a percent, you can easily change a percent to a decimal.

Whenever a number is written as a percent, it can be changed to a decimal. The rule is: *Move the decimal point two places to the left, and remove the % symbol.*

What if you do not see a decimal point?

Consider the following percents: 7%, 12%, and 175%. As you may recall, whole numbers have an *understood* decimal point to the *right* of their last numeral. Thus 7% = 7.0%; 12% = 12.0%, and 75% = 75.0%.

Let us change these percents to decimals. Move the decimal point *two* places to the left.

$$7.0\% = .07$$
$$12.\% = .12$$
$$175.0\% = 1.75$$

Now, change the following percent to decimals:

$$37.5\%, \ 0.05\%, \ 2.05\%$$

Move the decimal point *two* places to the left:

$$37.5\% = 0.375$$
$$0.05\% = 0.0005$$
$$2.05\% = 0.0205$$

A quick review: When changing from decimal to percent, move the decimal point two places to the right, and put in the % sign. When changing from percent to decimal, move the decimal point two places to the left, and remove the % symbol.

Chapter 7
Percent, Fraction, and Decimal Equivalents

When we work with percent applications in our everyday math problems, we will be using percents, and sometimes changing them to fractions or decimals. Here are tables of some important percents with their fractional and decimal equivalents. When someone says "50%" you see in the table that it represents the fraction 1/2 or the decimal 0.50.

Percent	Decimal	Fraction
100%	1.00	1
75%	0.75	3/4
50%	0.50	1/2
25%	0.25	1/4
1%	0.01	1/100
5%	0.05	5/100*
105%	1.05	1 5/100*
15%	0.15	15/100*
33 1/3%[1]	0.33333...**	1/3
66 2/3%[2]	0.66666...**	2/3

*These fractions are not in their "lowest terms." That is, 5/100 can also be written as 1/20 because they are equivalent. The fraction 15/100 is equivalent to 3/20.
** These numbers are called "continuing decimals." We can round, or shorten them so that 0.3333... is approximately 0.33, and 0.6666... is approximately 0.667, etc.
[1] Note that 1/3 = 0.333...   [2] Note that 2/3 = 0.666..

Here is another table of percents and decimal equivalents. This table has very useful percents that we use in our everyday lives such as sales taxes and stocks and bonds. There are state sales taxes of 6 3/4% or 7½% and the like. In addition, it shows the equivalency of *fractional* percents! That is, you sometimes hear "one-half of one percent" which is 1/2%.

Percent	Decimal
8 1/4%	0.0825
7 1/2%	0.075
6 3/4%	0.0675
1/2%[1]	0.005
1/4%[2]	0.0025
2 1/2%	0.025
0.6%[3]	0.006
6.25%[4]	0.0625

[1] 1/2% is also read as "one-half of one percent."
[2] 1/4% is read as "one-quarter of one percent"
[3] 0.6% is read as "six tenths of one percent."
[4] 6.25% is the fraction 1/16 (change 1/16 to a decimal on your calculator, the display will read 0.0625).

There are other percents such as 15.625% and 10.9375%. The percent 15.625% has a decimal equivalency of 0.15625 and is equal to the fraction 5/32 (a useful fraction on the stock market). (If you start with the fraction 5/32 and change it to a decimal by dividing the denominator, 32, into the numerator, 5, you will get the quotient, the decimal equivalent 0.15625.) The percent 10.9375% has a decimal equivalency of 0.109375 and is equal to the fraction 7/64.

One-half of one percent is written as 1/2%, which is equal to 0.005, and that decimal can be changed to a fraction and read as "five-thousandths," which reduces to:

$$\frac{5}{1000} = \frac{1}{200}$$

"Mom," asked Marie, "I'm late for my classes. Would you please do my math homework for me?"

Marie's mother was annoyed. She wanted her daughter to do her own work! So she said, "Marie, it just wouldn't be right!"

"Well," said the daughter, "at least you could try!"

Chapter 8
Ratios and Proportions

A ratio is a comparison of one number to another. The ratio of A to B can be written as A:B or A ÷ B or A/B, where B can never be zero. The ratio, which is a fraction, should be reduced to lowest terms.

We can write the ratio of men to women in a computer class where there are 20 men and 16 women. Write the ratio as 20:16, or 20/16, which reduces to 5/4 as follows:

$$\frac{20}{16} = \frac{20 \div 4}{16 \div 4} = \frac{5}{4}$$

The ratio of 125 sunny days to 90 rainy days is written as 125:90 or 125/90, which reduces to 25/18.

$$\frac{125}{90} = \frac{125 \div 5}{90 \div 5} = \frac{25}{18}$$

Since ratios are the comparison of two numbers, we do not change the resulting fraction 25/18 to a mixed number. The ratio is read as: "25 to 18."

Can you think of other applications of ratios (besides odds in betting)? My water company sends me information about the health related standards in my drinking water. For radioactive substances, they list the amounts as *pico curies per liter*; for other substances they list *milligrams per liter*. These are ratios: so many milligrams for every liter of water, or milligrams/liter.

When we have two equal ratios, we have a proportion. A proportion may be written as:

$$A/B = C/D \quad \text{or} \quad A:B = C:D$$

where B and D are not zero. (Note: B and D are the denominators of the fractions, and denominators of fractions can never be zero.)

When are two fractions equal? When they are equivalent! *Fractions are equivalent when they represent the same amount.* So that, proportions are just another way of writing two equivalent fractions.

For example: 3/6 = 4/8 = 1/2. We have families of equivalent fractions: 2/4 = 3/6 = 4/8 = 5/10 = 6/12 = 7/14 = 8/16 = 9/18 = ½. This is demonstrated in the following pizza pies.

These pizza pies show that you can eat 3 pieces from a pie that has been cut into 6 equal pieces, or you can eat 4 pieces from the same size pie that has been cut into 8 equal pieces and you will have eaten the same amount!

Now consider these equivalent fractions:

3/9 = 6/18 = 12/36 = 24/72 = 48/144 = 96/288 = 1/3

These pizza pies show that you can eat 3 pieces from a pie that has been cut into 9 equal pieces, or even eat 6 pieces from a pie that has been cut into 18 equal pieces and you will have eaten the same amount!

Remember, proportions are two equivalent fractions. In the proportion:

$$\frac{A}{B} = \frac{C}{D}$$

the values of A and D are called the *extremes* of the proportion. (Notice that A and D are at the extreme positions of the proportion.) The values of B and C are called the *means* of the proportion. (Notice that B and C are in the middle positions of the proportion.) You can also write a proportion as:

A:B::C:D

which can be read as: A is to B as C is to D. In this form, you can see that A and D are the extremes, and that B and C are the middle values, the means.

In a proportion, the product of the means is equal to the product of the extremes.

In the proportion 3/6 = 4/8, the means are 6 and 4 and the extremes are 3 and 8. The product of the means is equal to the product of the extremes: $3 \cdot 8 = 6 \cdot 4 = 24$.

Chapter 9
Solving Proportions

What if...

...a famous Chef gave us his recipe for a chocolate souffle serving 80 people, and we had a small gathering of 8 friends for dessert? How would we be able to use the chef's recipe?

...a nurse had to give certain medicine to a patient and the dosage was to be determined by the weight of the patient?

These preceding scenarios are some of the many applications for proportions.

When three of the four values in a proportion are given, we would like to solve for the fourth, or unknown, value.

To solve a proportion (two equal fractions) means to find the one missing value.

Here is a powerful technique to use when one of the four numbers of a proportion is unknown.

Suppose our proportion is:

$$\frac{?}{6} = \frac{4}{8}$$

We are going to *solve* this proportion. This means we will find the value of the unknown quantity represented here as a question mark (?). To solve the proportion, use the property that the product of the means is equal to the product of the extremes. Some say that we *cross multiply*, because that is what is happening in the proportion.

Designate the missing number, now shown as ?, as a letter, for example X. This quantity, X, as an unknown, is also called a *variable*. Now our proportion is:

$$\frac{X}{6} = \frac{4}{8}$$

The means are 4 and 6, and the extremes are 8 and X. We can write $8 \cdot X = 6 \cdot 4$. This is called an *equation*.

Notice that this equation has an *equal* sign (=). The left side of this equation equals the right side of the equation. (Any operation worked on one side of the equation, such as addition, subtraction, multiplication, or division, must also be done on the other side of the equation to maintain the correct equation balance.)

Our equation reads as follows: 8 times X is equal to 6 times 4. In other words, 8 times what number is equal to 24? You know that the answer is 3, so X = 3.

Note: In this *equation*, we can also write: $8X = 24$ (without the multiplication symbol), because it is understood that when there is a letter (a variable) multiplying a number the multiplication dot ($\cdot$) is not needed. Also, note that the number 8 that multiplies X is called the *coefficient* of X. If the coefficient of a variable (for example, X) is 1, we do not need to write the coefficient 1, it is understood to be 1. Just write $1 \cdot X$ as X.

Without guessing, we can find X as follows: divide *both* sides of the equation by the number multiplying the variable X (in this case, 8). 8 X = 24, 8/8 X = 24/8 (remember that 8/8 = 1), 1·X = 3, therefore X = 3.

In *any proportion*, A/B = C/D, if we know the value of three of the variables, we can always find the fourth unknown variable.

Let's get back to that famous chef and his "chocolate souffle" recipe. His recipe for 80 people calls for 20 cups of milk and 1 pound and 4 ounces (that's 16 ounces + 4 ounces = 20 ounces altogether) of chocolate, among other ingredients.

We can set up a proportion because the ingredients must be in the same ratio as the number of people to make the same kind of souffle. That is:

$$\frac{20 \; Cups \; of \; milk}{X \; Cups \; of \; milk} = \frac{80 \; People}{8 \; People}$$

The product of the means is equal to the product of the extremes. The means are 80 and X, the extremes are 20 and 8. When you cross multiply, you set these products equal to each other.

$$80X = 20 \cdot 8$$

$$80X = 160$$

$$\frac{80X}{80} = \frac{160}{80}$$

$$X = 2 \; cups \; of \; milk.$$

What about the chocolate amounts? Here is the proportion.

$$\frac{20 \ Ounces \ of \ chocolate}{X \ Ounces \ of \ chocolate} = \frac{80 \ People}{8 \ People}$$

Cross multiply and solve the equation: 80X = 160; therefore, X = 2 ounces of chocolate needed.

This proportion could also have been written as:

$$\frac{20 \ Ounces \ of \ chocolate}{80 \ People} = \frac{X \ Ounces \ of \ chocolate}{8 \ People}$$

Notice that if you cross multiply in this case, you will get the same equation as before, so that you would get the same answer: 20 · 8 = 160; 80X = 160, X = 2 ounces of chocolate.

What about that nurse who had medicine to be administered at a dosage of 3 ounces for every 150 pounds of weight? The patient was a very young boy who weighed only 50 pounds. How much medicine should he be given? Set up a proportion:

$$\frac{3 \ Ounces \ of \ medicine}{X \ Ounces \ of \ medicine} = \frac{150 \ Pounds}{50 \ Pounds}$$

The product of the means is 150X and the product of the extremes is 3 · 50. When you cross multiply you set these products equal to each other, as shown:

$$150X = 3 \cdot 50$$

$$150X = 150$$

$$\frac{150X}{150} = \frac{150}{150}$$

X = 1 ounce of medicine.

The little boy will get 1 ounce of the medicine. In these medicinal problems, it can be a life and death matter; too much medicine or not enough medicine can be dangerous!

Did you notice that the problems had *like items* on each side of the proportion: milk/milk = people/people and medicine/medicine = pounds/pounds.

Another application for ratios and proportions is scale drawings such as maps (drawn to scale). This application will be discussed in Part 2.

And, if you are a world traveler you can definitely use ratios and proportions when it comes to shopping in other countries. You would have to know how to change (or convert) dollars to whatever currency is used in the foreign country you are visiting.

There you are in England, and you want to buy a sweater that costs 92£. (This means 92 English pounds.) So, how many dollars is that?

You will find this to be an interesting task; however, you may want to carry a small hand calculator with you when you travel!

Let us assume that the conversion rate is 1£ = $1.50 at the time of this transaction.

This means that one English pound (£) is worth $1.50 U.S. The sweater costs 92£. So, we set up a proportion as follows:

$$\frac{1 \; English \; Pound \; (£)}{1.50 \; U.S. \; Dollars (\$)} = \frac{92 \; English \; Pounds}{X \; U.S. \; Dollars}$$

The conversion rate is the left side of this proportion. Cross multiply, and solve for X:

$$X = (1.50) \cdot (92) = \$138.$$

This English sweater, therefore, costs $138. Later, when we work with percents, you will see that you will also learn how to figure out the VAT (Value Added Tax).

Notice that the left side of the proportion showed the conversion of £ to $. We must keep the same ratio of £ to $ on the right side.

Now, we will travel to country ZZZ where the U.S. dollar is worth 50,000☺. We would like to buy a sweater that is selling for 2,750,000☺. How much is that in U.S. currency?

As long as you have the conversion rate, you can write the proportion and convert any currency.

Set the proportion up like this:

$$\frac{\$1}{50,000☺} = \frac{\$X}{2,7500,000☺}$$

Cross multiply so that: 50,000X = 2,750,000, and X = 55 U.S. dollars.

Where did we get the conversion rate? Banks or money changers in various countries will give you the rates. Currency conversion rates are also published daily in newspapers.

Many countries use the Metric System, so besides having to convert the currency, travelers may also have to change kilograms into pounds and kilometers into miles. Here are some examples of converting from and to metric system:

1. There are some nice fresh apples in the supermarket. Each bag of apples weighs 5 kilograms. How many pounds is that?

2. If I only want 1 pound of green beans, how many kilograms should I ask for?

3. You drive along a highway and see a speed limit sign that says: 110 km/hr. How fast is that in miles/hour?

We really need to remember just a few metric conversions to get around. That is:

    1 kilogram ~ 2.2 pounds (lbs)
    1 kilometer ~ 0.6 miles

We can use proportions to help us find the conversions for the metric values.

1. Set up the proportion for the apple problem as follows:

$$\frac{1\ Kilogram}{2.2\ Pounds} = \frac{5\ Kilograms}{X\ Pounds}$$

Cross multiply: X = 5 · 2.2 = 11.0 pounds. This tells us that 5 kilograms of apples weighs 11 pounds.

2. Set up the proportion for the green bean problem as follows:

$$\frac{1\ Kilogram}{2.2\ Pounds} = \frac{X\ Kilograms}{1\ Pound}$$

Cross multiply. This time we have: 2.2X = 1.0. Divide 1.0 by 2.2 to get the answer: X = 0.45 kilograms. This tells us that 0.45 kilograms weighs approximately one pound.

3. Now, set up the proportion for the speed limit sign:

$$\frac{1\ Kilometer}{0.6\ Miles} = \frac{110\ Kilometers/Hr}{X\ Miles/Hr}$$

Cross multiply: X = 110 x (.06) = 66 miles/hr.

So, dear reader, when you see the 110 Km/Hr speed limit sign on the highways in different countries, don't drive 110 miles per hour! The speed limit is really 66 miles/hr.

Similarly, if someone from a country that is on the Metric System comes to visit you and sees a speed limit sign of 45 miles/hour, how can you be helpful and find the conversion into kilometers/hour?

Set up the ratio and proportion as follows, then cross multiply to find the kilometers/hour:

$$\frac{1\ Kilometer}{0.6\ Miles} = \frac{X\ Kilometers/Hr}{45\ Miles/Hr}$$

Cross multiply this proportion so that 0.6X = 45. Divide 45 by 0.6 (you may use your hand calculator). Thus, we find that X is 75 kilometers/hour. Here is a shortcut converting from kilometers to miles: multiply the number of kilometers by 0.6. For example, 75 kilometers/hr x 0.6 = exactly 45 miles/hour!

Chapter 10
The Percent Proportion

How would you know...
...the best deal (lowest cost) if you need to borrow money?
...that you received the correct discount at a store sale?
...how much your lawyer received in fees if he told you he was on a 33% retainer?
...that your medical insurance paid the contracted amount (sometimes 80%) of the hospital charges?
...what those deductions on your payroll stub are and if the correct amounts been deducted?

I will show you how to answer the above questions and solve proportions. This leads us to the *percent proportion*. This will be used to solve *all* the percent problems to be presented to you. The percent proportion is:

$$A/B = P/100$$
or
$$\text{Amount/ Base} = \text{Percent/100}$$

To understand this percent proportion, recall that percent means *parts per hundred*. If you write 30%, that represents *30 parts per hundred*, or 30/100 as a fraction. P% represents P parts per hundred, or P/100. The right side of the percent proportion is the percent, written as P/100.

The left side of the percent proportion is a *ratio* of the amount, A, to the base, B. The amount and base are related because the amount is a certain portion of the base. When this fraction (the ratio A/B) is set equal to the percent fraction P/100, you have the percent proportion. (You have two equal fractions.)

To solve this percent proportion, we must be given any *two* of the values A, B, or P. Thus, there are *three* types of equations, in which any one of the three variables A, B, or P may be the unknown.

Here is a special method that may help you understand how to use the percent proportion. In any given problem you will have the values for two of the three variables A, B, or P. How can you determine from the problem, which value represents the amount A, or the base B? You ask the following question:

What percent P *of* B is A?

If you are given any type of percent word problem, you can use the above question as follows: identify the percent (P) because it will be the number followed by the percent sign %. Identify the base (B), which is the total collection of items for the problem because it is usually the number following the word *of* or *out of*, or some other way of expressing the total sample in your problem. The amount (A) sometimes follows the word *is*.

Example 10-1. What percent of 5 is 2?

This illustrates the question: What percent (P) of base (B) is the amount (A)?

You see that the percent is the unknown value. The value of the base is 5, and the amount is 2. Now you can set up the percent proportion and solve for the percent as follows.

$$\frac{A}{B} = \frac{P}{100} \text{ or } \frac{2}{5} = \frac{P}{100}$$

Cross multiply this proportion, so that:

$$5 \cdot P = 2 \cdot 100$$
$$5P = 200$$
$$P = 40$$

Thus, the answer is 40%.

Example 10-2. 25% of what number is 50?

In this example, the percent (P) has the value 25. The phrase "of what number" shows that the missing value is the base (B). The value of the amount (A) is 50.

You set up the percent proportion like so:

$$\frac{50}{B} = \frac{25}{100}$$

Cross multiply this proportion so that you have:

$$25 \cdot B = 50 \cdot 100$$
$$25B = 5000$$
$$B = 200$$

Thus, we find B to be 200. You can check this answer. Is 25% of 200 equal to 50? (Yes.)

Now that you understand the way the percent proportion works, I will show you two shortcut methods to solve certain percent problems. These shortcuts can be used to find the amount (A) or the percent (P). When the base (B) is the unknown, you work the problem as shown above.

I would also like to explain how to use a hand calculator to solve these percent problems.

Example 10-3. 10% of 130 is what number?

In this example, you are given the percent (P) as well as the base (B). The missing value is the amount (A). Set up the percent proportion as in the previous examples:

$$\frac{A}{130} = \frac{10}{100}$$

Cross multiply this proportion so that:

$$100 \cdot A = 10 \cdot 130$$
$$100A = 1300$$
$$A = 13$$

Therefore, 10% of 130 is 13.

These three examples illustrated all three possibilities using the percent proportion.

I would also like to explain how to use a hand calculator to solve these percent problems.

Shortcut #1. If you are given the base (B) and the percent (P), then this shortcut method is used to find the amount (A). Change the percent to a decimal. (Remember, percent is parts per hundred. To change a percent to a decimal, remove the % sign, and move the decimal point *two places to the left*. If the percent is a whole number, there is an assumed decimal point after the right-most digit.) Once you change the percent to a decimal number, multiply by the base (B) and watch those decimal points!

This shortcut works very well on a hand calculator after you have changed the percent to a decimal. You multiply the decimal value of (P) by the base (B), and the display will show the correct amount (A).

If you happen to have a hand calculator with a % key, then key in the value of the base (B), press the multiply key, then key in the value of the percent (P) and then press the % key. Your answer for the amount (A) should be in the display. The % key saves you the time of changing the percent to a decimal!

Remember, this % key *does not change the percent to a decimal for you.* However, when you key in the base, and press the multiply key, then key in the value of P, then press the % key, the multiplication will be done correctly, and the value of A, the amount, will be in the display.

Shortcut #2. This is a shortcut method for finding the percent (P) when the amount (A) and the base (B) are known values.

Form the fraction A/B. Change the fraction A/B to a decimal by dividing: A ÷ B. Then change the decimal to a percent by moving the decimal point two places to the right, and putting in the % sign.

This shortcut to find (P) if (A) and (B) are known also works very well on a hand calculator. Key in A into the hand calculator, then press the ÷ key. Key in B and press =. The fraction A/B has been calculated, and appears in the display as a decimal number. Now, change the decimal to a percent as stated above.

If your hand calculator has a % key, you can find the percent, P, as follows: key in the amount, A, press the ÷ key. Key in the value for the base, B, then press the % key. The *value* for the percent appears in the display. However, there is *no* % sign behind the number, you must fill that in yourself. You can refer to these methods as you study the applications throughout the rest of the book.

Now, back to the questions at the start of this chapter.

If you need to borrow money, you know that there will be interest due and payable on any loans. The base, is the amount of money to be borrowed, and the rate of interest is the percent P. You can use shortcut #1, multiply the rate (change to a decimal) by the amount of money borrowed. The answer is the yearly interest due.

To find the best deal you would have to compare different rates, use shortcut #1 to find the various amounts and make your wise decision. Here is an example of finding the interest due on a loan. You can use the percent proportion or the shortcut #1 method, since the missing value is the amount, A. I will work this example both ways.

Example 10-4. Interest on a Loan.

You borrow $6,000 at an annual interest rate of 8%. What is the amount of interest per year that you would be required to pay?

A is the amount of annual interest, B is the original loan amount, and P is the interest rate. Thus, our percent proportion becomes:

$$\frac{Amount\ Of\ Annual\ Interest}{Original\ Loan\ Amount} = \frac{Annual\ Interest\ Rate}{100}$$

The original loan amount is $6,000. The annual interest rate is 8%. Thus, B = 6000, P = 8 and we have:

$$\frac{A}{6000} = \frac{8}{100}$$

In this proportion, solve for the unknown variable amount A using: "The product of the means is equal to the product of the extremes," or cross multiply as follows:

$$100A = 48,000$$

$$then \ \frac{100A}{100} = \frac{48,000}{100}$$

so that A = $480, the interest required.

Now, use the shortcut #1 to find the amount (A) as follows: change 8% to a decimal: 0.08, then multiply that decimal by the base (B): $0.08 \cdot \$6,000 = \$480$. I will show you this method whenever we solve for the unknown (A).

If you use a hand calculator with a % key, you key in 6000, press the × (multiplication) key, then key in 8, press the % key. The answer 480 should be in your display.

Example 10-5. Percent discount problem. Lisa bought a TV. The dealer wanted $760 (list price) but Lisa got a discount of 25% off the list price. How much was the discount? How much did Lisa pay for the TV? The list price is B, $760. The discount, A is unknown; P is 25%.

$$\frac{A}{760} = \frac{25}{100}$$

Cross multiply, solve the percent proportion:

$$100A = 19000$$

$$then \ \frac{100A}{100} = \frac{19000}{100}$$

so that A = \$190: amount of the discount.

Let us use shortcut #1 in this problem, because the amount is the unknown value. The discount of 25% = 0.25. Then, multiply it by the list price: 0.25 x \$760 = \$190, the amount of the discount.

Thus, Lisa paid \$760 - \$190 = \$570 for the TV.

Because Lisa received a 25% discount, she actually paid: 100% - 25% = 75% of the list price for the TV. Now, P is 75% and B is \$760. Thus, we can use the shortcut method directly to find the amount Lisa paid: 75% = 0.75; 0.75 x 760 = \$570.

Notice here that I have written: 100% - 25% = 75%. Yes, you can add, subtract, multiply, and divide percents. Since they can be changed into decimals, you can do the arithmetic with them. When you work this problem, and work with 75% of the list price, you are finding the discounted price immediately.

Example 10-6. How would we know how much our lawyer received in fees if she told us she was on a 30% retainer? Here is an application of lawyer's fees.

Colette's ex-husband owes her \$25,000 in child support. Colette's attorney told her she could successfully sue, for a fee of 30% of what they collect from the ex-husband. If Colette agrees to the terms, and is awarded $25,000, how much does she net?

In this percent problem, the lawyer's fee is the percent, 30%, and the base, is $25,000. You can use the shortcut method to find the amount. Change the percent to a decimal and multiply by the base: 30% = 0.30, 0.30 x $25,000 = $7,500. This is what the lawyer charged. Colette receives: $25,000 - $7,500 = $17,500.

Another way of looking at this problem is to say that Colette would receive 100% - 30% = 70% of the settlement. Now, P is 70% and B is still $25,000. 70% = 0.70, 0.70 x $25,000 = $17,500.

Chapter 11
Percent Problems and Their Solutions

Now it is time to work some additional percent applications. There are many types of everyday percent problems, and you can look them over in any order you choose, whatever is more meaningful for you. I am grouping these problems by their type. You will see the following types: commissions, loan interest, discounts, percent passing or failing, sales taxes, percent decrease or increase, profit or loss on stock market, some assorted percent problems, and, last but not least, income tax problems.

In all these problems, I encourage you to use the shortcut methods, and I show step-by-step methods to work hand calculators. Naturally, you'll want to make sure the input data is correct when you do any calculations.

Joe: Tell me, what is worse than finding 100% of a worm in your apple?

Moe: I know, finding 50% of the worm in my apple!

Type 1: Commissions

Example 11-1. Real Estate Commissions.

Jan is a real estate broker who sold the Jones their $190,000 house. She earned a 6% commission on the sale. How much money did Jan earn?

In this example, the original price of the house is $190,000, and P = 6% commission rate. The amount of commission dollars is the unknown. We can use shortcut method #1, which works when A is the missing quantity.

To solve this problem, change 6% to a decimal, then multiply the decimal by the base. 6% = 0.06, and 0.06 x 190,000 = $11,400.

Jan earned $11,400 as her commission.

Example 11-2. A Salesperson's Salary and Commission.

Peggy is a salesperson and has a salary of $250/week plus a 30% commission on what she sells over $1,200 per week.

A. Last week, her sales amounted to $2,800. How much did Peggy earn that week?

First, find the amount that is more than $1,200. So that: $2,800 - $1,200 = $1,600. Peggy earns a commission of 30% on this amount, which is the base, B. The commission rate is the percent, P which is 30%. We can solve for A using the shortcut method: Change 30% to a decimal and multiply by the base: 30% = 0.30; 0.30 · 1,600 = $480 commission earned.

Peggy earned $250/week plus this amount of $480, so that she earned: $250 + $480 = $730.

B. How much must Peggy sell per week to maintain a weekly income (salary + commission) of $1,000?

We know that Peggy receives a weekly salary of $250. She needs to earn: $1,000 - $250 = $750 in comm-issions.

The total amount of sales, B, is unknown. The amount, A, is $750 and P is 30%. We cannot use the shortcut methods here.

Let us set up a percent proportion and cross multiply to find the total amount of sales Peggy needs.

$$\frac{750}{B} = \frac{30}{100}$$

$$30B = 75,000$$

$$B = \$2,500$$

B = $2,500, the amount of sales *over $1,200* needed by Peggy to maintain her weekly income of $1,000.

Thus Peggy needs to have sales of $3,700/week ($2,500 + $1,200) to maintain a weekly income of $1,000.

C. Let us check this out. First, find the amount that is over $1,200. Thus: $3,700 - 1,200 = $2,500.

Peggy earns a commission of 30% on this amount. Note here that $2,500 is the base and P is 30%. We can solve for A using the shortcut method. Change 30% to a decimal and multiply by the base:

$$30\% = 0.30; \quad 0.30 \cdot 2,500 = \$750.00$$

Peggy earns $250/week plus $750 in commissions, so that she earned $250 + $750 = $1000.

Example 11-3. Royalty Percents.

Jim wrote a book, and the publisher agreed to pay him a royalty of 10% of the net price.

The retail price of the book is $24. The stores buy the book at net price from the publisher, and then mark up the price 20% to get their retail price.

Jim wants to know how many books must be sold for him to earn $1,500.

First, Dear Reader, we have to find the net price of the book. The retail price is $24, which includes the 20% markup. Consider: 100% of the net price + 20% of the net price = retail price. So that: 100% + 20% = 120% of net price is the retail price.

Can we find the net price? P is 120%, the base, B, is the net price, and the amount, A, is the retail price, which is $24.

Set up the percent proportion, cross multiply to find the base (B):

$$\frac{\$24}{B} = \frac{120}{100}$$

$$120B = \$2,400$$

$$B = \$20 \ \textit{Net price}$$

Jim gets 10% royalty on the net price of each book. Now that we know the net price is $20, we can use the shortcut method to find how much Jim earns per book: P is 10%, the base, B is $20.

Find A by multiplying the percent (changed to a decimal) by B: 10% · $20 = 0.10 · $20 = $2 per book.

So for Jim to earn $1,500, the bookstores will have to sell $1,500 ÷ $2 = 750 books. (Earnings ÷ earnings per book = number of books.)

Let's hope you make the best seller list, Jim!

Joe: "Hey, Moe, bet you don't know why skunks do so well in math calculations?"

Moe: "Tell me why."

Joe: "Skunks have great PER SCENTS!!"

Type 2: Interest

Example 11-4. Banks and Savings Institutions.

These institutions pay interest on deposited funds. Let us say they are paying a 6% simple interest annual rate.

Lois deposits $900 into her savings account for three months (that is, one-quarter of a year). How much simple interest will her deposit earn?

First, let us find the *quarterly* rate, which is 1/4 of 6%. (You can do the arithmetic with percents, 1/4 of 6% = 1.5%.) Use this new 1.5% to find the interest Lois receives. P is 1.5%, the base (B) is $900, and the amount of interest, A, is the unknown. We can use the shortcut method to find A. Change the percent P to a decimal and multiply by the base. P = 1.5% = 0.015 as a decimal. (Remember, we move the decimal two places to the left and remove the % sign.) 0.015 · $900 = $13.50 for the quarter of the year.

If Lois leaves the $900 in the bank for the whole year (at simple interest), you find how much simple interest she earned by *multiplying* the amount of interest earned in the quarter by 4 (4 quarters in a year). $13.50 · 4 = $54.00 for the year. Notice: $900 deposited for a year at 6% is $900 · 0.06 = $54.00.

Example 11-5. Bank Credit Cards and Charges.

Bank XX issues credit cards that have interest charges on unpaid balances as follows. If the average daily balance is $500 or less, the monthly interest is 1% of the balance due. If the average daily balance is over $500, the monthly interest is $5 plus 3/4% of the balance due that is greater than $500. There are no annual fees, nor is there any monthly interest if the customer pays the full account balance monthly.

Brian has a Bank XX credit card and has an unpaid average balance of $850. How much interest does Brian have to pay this month on his unpaid balance?

According to his bank's calculations, Brian should pay $5 plus 3/4% of the amount (of the unpaid balance) over $500. His unpaid average balance is $850. Thus: $850 - $500 = $350. Brian should pay $5 plus 3/4% of $350.

We can use the shortcut method to find 3/4% of $350. P is 3/4%, and B is $350.

What does 3/4% mean? Change 3/4 to a decimal: 3/4 = 0.75, so that, 3/4% = 0.75%. Now change 0.75% to a decimal by moving the decimal point two places to the left and removing the % sign: 0.75% = 0.0075.

Now use the shortcut method: 0.0075 · $350 = $2.63. Brian's interest this month on his unpaid average daily balance is $5 + $2.63 = $7.63.

Example 11-6. Interest on loans.

Carmen bought a used car for $7,000 less a 10% discount. She paid 25% of the agreed-upon amount immediately, and took a loan for the balance. The loan had an annual rate of 10% simple interest.

1. How much did Carmen pay up-front (in other words, how much money did she put down for the car)?
2. How much did she have to borrow for the car?
3. How much is the (yearly) amount of the simple interest on her loan?

The 10% discount on the price of the car, $7,000,

can be obtained by multiplying: 0.10 x $7,000 = $700 discount. This means that Carmen agreed to buy the car for $7,000 - $700 = $6,300. When she paid 25% of this amount up-front, she paid 25% of $6300. The amount, A, is the unknown value, so we use the shortcut method to find A.

1. 25% = 0.25, 0.25 · 6300 = $1575 paid up-front.

2. Carmen has to borrow the rest, or $6300 - $1575 = $4725. Therefore, the loan amount is $4725, the percent interest on the car loan is 10%, and the unknown value is A, her amount of yearly interest.

3. We can use the shortcut method to find the Amount A. The Percent is 10%, so that: 10% = 0.10; 0.10 · 4725 = $472.50 interest per year.

Type 3: Discounts

Example 11-7. Fred purchased a used car that had a list price (asking price) of $12,000. The car dealer gave Fred a $3,000 discount, so that Fred paid $9,000 for the car. What was the percent discount?

In this example, the base (list price) is $12,000 and the discount is $3,000. We can use shortcut method #2 to find the percent discount as follows: Set up the fraction 3000/120000 (enter 3000 press the ÷ key, enter 12000 press =). Change this amount into a percent: the decimal 0.25 = 25%.

The percent discount for Fred was 25%.

You can check this problem: 25% off the price of the used car, which is $12,000 is found by shortcut method #1, where P is 25%, and B is $12,000. Change 25% to a decimal and multiply: 25% = 0.25, 0.25 x $12,000 = $3,000. That was the discount the dealer gave Fred.

You can use your hand calculator for this check: Key in 12,000, press x then key in 0.25 and press =. If you have the % key, key in 12,000, press x, key in 25, and press %. The display will read 3,000.

Example 11-8. Here is another percent problem involving a discount on a car. The used car dealer offered a discount of 20% off the price of a certain car. The discount was $2,000. How much was the *original price* of the car?

We are given the percent (P = 20). The base would be the original price, and is unknown. The amount is the discount of $2,000. Put these facts all together, and solve the proportion:

$$\frac{2,000}{B} = \frac{20}{100}$$

Cross multiply, solve the percent proportion:

$$20B = 200,000$$

$$\textit{then } \frac{20B}{20} = \frac{200,000}{20}$$

so that B = $10,000, *the original price of the car.*

We can check this problem as follows: If the original price of the car is $10,000, and the discount is 20%, then the amount of the discount is found using the shortcut method to find A. Change the 20% to a decimal, multiply by the base: 20% = 0.20, therefore: 0.20 · $10,000 = $2,000, which was the discount as stated above.

Type 4: Students Passing and Failing

Example 11-9. This problem involves finding the percent of students passing exams. In a nursing class, 25 students took the final exam, and 20 students passed. What percent of the students passed?

The total amount, B, is 25. The number passing the exam, A, is 20. Find the percent, P, of those passing.

You can use shortcut method #2 to find the percent. Form the fraction A/B and change the resulting decimal answer to a percent.

$$20/25 = 0.80 = 80\%.$$

If your hand calculator has a % key, you can work this problem as follows: key in 20, press ÷, key in 25, then press the % key. The value of the percent (without the % symbol) will be in your calculator's display.

If you know that 80% passed, then you also know that 100% - 80% = 20% of the students failed.

Example 11-10. In a math class, the final grades were as follows: 6 students received As, 10 students received Bs, 18 students received Cs, and 6 students failed the course. What percent of the class passed the course? That is, what percent of the class received A or B or C grades? What percent failed the course?

The total number of students is found by adding all the students who received any grade at all: Thus B, the base = 6 + 10 + 18 + 6 = 40. The total number of students who passed the class is found by adding all the A, B, and C students: 6 + 10 + 18 = 34.

You can use shortcut method #2 to find the percent

of students passing the course. Form the fraction 34/40. Key in 34 in your hand calculator, press ÷ then key in 40, press =. The display will show the decimal: 0.85. Now, change the decimal to a percent (move the decimal point two places to the right, and put in a % sign). 0.85 = 85% passed the course.

There were 6 students who failed the course. In like manner you can use shortcut method #2 to find the percent of students failing. Form the fraction 6/40. You find the decimal result to be 0.15 or 15%.

The percent of students passing is 85%, the percent of students failing is 15%. Add 85% + 15% = 100%. This corresponds to the whole class of 40 students, since 100% represents the whole class. In fact, 100% - 85% = 15%. The whole class - those passed = those failed.

Example 11-11. Here is a percent problem involving test taking. This year, 85% of those taking the bar exam (lawyers' exams) passed.

1. If 2,720 people passed the exam, how many took the exam? (Hint: we are looking for the total number, the base.)
2. What percent of those taking the bar exam this year failed?
3. How many people failed the exam?

1. The total population is unknown, B. The amount of people passing the exam, 2,720, is A, and the percent passing is 85%. Solve the percent proportion for B.

$$\frac{2,720}{B} = \frac{85}{100}$$

Cross multiply to solve the percent proportion:

$$85B = 272,000$$

$$then \ \frac{85B}{85} = \frac{272,000}{85}$$

so that $B = 3,200$ *total number taking exam.*

2.   If 85% passed, then 100% - 85% = 15% failed the bar exam this year.

3. We can find out how many failed in two ways. First, 15% of the total number of students failed, or: 15% of 3200. Use the shortcut method, so that: 15% = 0.15; and $0.15 \cdot 3,200 = 480$ failed.

Thus, the total number of students - those who passed = those who failed. So that we have: 3,200 - 2,720 = those who failed, 480.

Type 5: Sales Taxes

Here is a topic that reaches into all our lives. It concerns sales tax on merchandise. Many times the sales tax is a mixed fraction with percents such as 8 3/4%, or 7½% and the like. You can refer back to our table on page 23 to see the fractional and decimal equivalence of these strange sales tax numbers. We work the problems in the same manner as before. It helps here to have a good hand calculator, and carefully enter the data.

Example 11-12. This is a percent problem involving sales tax. If the state sales tax is 8 3/4%, how much tax did Colleen have to pay when she bought her $13,500 car?

The percent, P, is 8 3/4%, the total purchase price, B, is $13,500, and the amount of the tax, A, is unknown. Notice that we can write 8 3/4% as 8.75%.

This problem can be worked using shortcut method #1, because A is the missing variable. 8 3/4% = 8.75% = 0.0875. Multiply 0.0875 × 13,500 = $1,181.25.

If your hand calculator has the % key, key in the actual percent, 8.75, press the x key, key in 13,500, then press the % key. The display will show 1,181.25.

Here is another sales tax example. Here I will show you how to find the total amount due (merchandise price plus tax), directly.

Example 11-13. Lisa bought a TV for $570. She also paid a 7% sales tax on that amount. We want to calculate the total amount Lisa paid for the TV. What was the amount of the sales tax? What was the total amount Lisa paid for the TV? The price, $570, is B. The sales tax of 7% is P, and the amount in dollars of the sales tax, is the unknown, A.

We calculate the sales tax, using shortcut method #1. Change 7% to the decimal 0.07. Then multiply 0.07 · 570. The sales tax amount is $39.90. The total amount Lisa paid for the TV set was the list price plus the tax: $570 + $39.90 = $609.90

Here is how you can get the total amount directly. You can add, subtract, multiply and divide percents (because they can be changed into their decimal form). Thus Lisa paid 100% of the list price, and then paid an additional 7% of the list price for the sales tax. You can write: 100% of list price + 7% of list price = 107% of list price.

Now, to find the total amount Lisa paid, we know the percent is 107%, the base is $570. Find the amount. Use shortcut method #1. 107% is the decimal 1.07. 107% = 1.07, 1.07 x $570 = $609.90.

The following is a variation on the previous sales tax problem.

Example 11-14. This percent problem involves sales taxes in a retail business. Marge owned a retail business. Her weekly sales record *included* the state sales tax of 7%. If her record showed sales of $22,684 for that week, how much did the merchandise sell for, *before* the tax was included?

In this problem, note that the weekly sales of $22,684 is equal to the original selling price *plus* the sales tax. The pretax selling price (which is the unknown in this problem) can be considered to be 100% of the sales. (Remember that 100% of anything is itself, or 100% = 1.)

Then, we reason that 100% sales + 7% sales (tax) adds up to $22,684 (weekly sales).

We can write: 100% + 7% = 107%. In this problem, the total percent (P) is 107%. The original selling price is the value of B, the unknown. A is the weekly sales including tax which is $22,684. In this example, we will have to set up the proportion to solve for B:

$$\frac{22684}{B} = \frac{107}{100}$$

Cross multiply and solve the percent proportion:

$$107B = 2268400$$

$$\textit{then } \frac{107B}{107} = \frac{2268400}{107}$$

so that B = $21,200, the pretax selling price.

You may wonder why this B is smaller than the value for A. This answer is consistent with the problem because B is the selling price *before* adding in the sales tax. In fact, let us check, by working backward, that the answers are correct.

If the pretax selling price is $21,200, and the tax is 7%, then the tax on this amount can be found by using the shortcut: 7% = 0.07, and 0.07 · 21200 = $1,484, sales tax. Add the tax to the selling price to obtain the weekly totals: $21,200 + $1,484 = $22,684.

Dear Reader, some of our examples illustrate how we can add or subtract percentages to simplify problems. Since percentages are a different form of decimal notation, we can do the arithmetic (add, subtract, multiply, divide) percents, as we do decimals. *Be sure you have the correct number of decimal places in your answer!* (You can check your answers on a hand calculator).

Here are some calculations you can do with percents:

10% + 25% = 0.10 + 0.25 = 0.35 = 35%
100% + 6 3/4% = 1.00 + 0.0675 = 1.0675 = 106 3/4%
100% - 15% = 1.00 - 0.15 = 0.85 = 85%
100% - 7 1/2 % = 1.00 - 0.075 = 0.925 = 92.5% (92 ½%)
10% *of* 25% = 10% x 25% = 0.10 x 0.25 = 0.025 =2.5%
15% *of* 80% = 15% x 80% = 0.15 x 0.80 = 0.120 = 12%
1/2 x 66% = 1/2 x 0.66 = 0.33 = 33%
36% ÷ 4 = 0.36 ÷ 4 = 0.09 = 9%

Type 6: Percent Increase and Decrease

You sometimes read or hear about percent increases or decreases. One of your friends may have received a pay increase of 5%, another had a 10% reduction in pay. Here are some examples to illustrate this.

Example 11-15. This is a problem involving percent increase in population. A census was recently taken and it reported that the population of City A was 500,000. Ten years ago, City A had a population of 400,000. What was the percent increase in the population?

In these increase or decrease problems, it is very important to find the correct base (B). The base (B) is the original (or the population 10 years ago) to which we compare the new one. The base B is 400,000. The increase in population is found using this expression:

new population- old population = increase

Thus: 500,000 - 400,000 = 100,000 population increase. This, then, is the value for the amount, A, in our percent proportion. We can set up the percent proportion to find P, and solve the proportion, or we can use shortcut #2, which is the following: Form the fraction A/B, change the resulting decimal to a percent. 100,000/400,000 = 1/4 = 0.25 = 25%.

Let us check our logic. The original population was 400,000. The percent increase was found to be 25%. How much was the population increase? The base (B) = 400,000 and the percent increase (P) = 25%. Find A using the shortcut method.

25 % = 0.25, thus 0.25 · 400,000 = 100,000

The population increased by 100,000. That is correct. The original population, 400,000 + the increase, 100,000 = 500,000, the present population.

Example 11-16. This is a problem involving percent decrease in salary. Carl was employed at a company and received wages of $48,000/year. For personal reasons, he decided to work part time for wages of $36,000. What is the percent decrease in his salary?

The base, B, is the *original amount* of his salary: $48,000. His decrease in wages is $12,000 ($48,000-$36,000). A decrease (or loss in business) is written with a - in front of the number or the number is written inside parentheses such as ($12,000). The decrease in wages is the amount, A. We solve for the percent, P. Note, we will use the - (minus sign) in front of the number to show the decreased amount.
Again, we can use shortcut method #2 to find P. Form the fraction A/B, change the decimal answer to a percent.
Thus we have: -12,000/48,000 = -0.25 =- 25%. This is a decrease of 25%.
Let's check the logic of the problem. Carl's original salary was $48,000. The Percent decrease was 25%. How much was the salary decrease? Find A by using the shortcut method. 25 % = 0.25, thus 0.25 · $48,000 = $12,000. Carl's salary decreased by $12,000. That is correct. The original salary, $48,000 less the decrease, $12,000 is equal to: $36,000, Carl's new salary.

Example 11-17. Are there any dieters here? It is always a pleasant feeling when the dieters find out they have lost some percent of their original weight. This percent problem involves percent loss. Jamie originally weighed 185 pounds. She dieted and lost 37 pounds. What percent of her original weight did she lose?

Jamie's original weight is B, 185 pounds. She lost 37 pounds, this is A. Find the percent, P. Here I show you the percent proportion, and how to solve for the value P using the percent proportion.

$$\frac{37}{185} = \frac{P}{100}$$

Cross multiply to solve the percent proportion:

$$185P = 3700$$

$$then \ \frac{185P}{185} = \frac{3700}{185}$$

$$so \ that \ P = 20\%.$$

That's pretty good dieting.

There are other examples of percent increases or decreases. For example, in buying clothes or other materials, sometimes there is a label on some items telling the consumer that there is a certain percent shrinkage. In the next two examples, I show the consumer how to find the amounts of shrinkage, which are really decreases in the size.

Example 11-18. Janetta went shopping and saw a great pair of jeans on sale. She tried on one pair, size: waist 27", length 30". They fit perfectly. The label inside read: "Guaranteed 3% maximum shrinkage." Should Janetta buy the jeans? If the jeans shrink the maximum 3%, what will the new measurements be?

We want to find the amount of shrinkage. Use the shortcut method. The value of P is 3%; the value of the base, B, is 27". Find the amount of shrinkage, A. Multiply the base by the decimal equivalent of the percent: 27" x 3% = 27" x 0.03 = 0.81". Thus, the waist will become, after shrinkage: 27" - 0.81" = 26.19".

The length will shrink, also. The value of P is 3%; the value of the base, B, is 30". Find the amount of shrinkage in the length: 30" x 3% = 30" x 0.03 = 0.90". Thus, after shrinkage, the length will become: 30" - 0.90" = 29.1".

We could have saved a step in the calculations. If the jeans shrink 3%, then their new size would be: 100% - 3% = 97% of the original size. Thus, we could have found the new waist and length sizes by using P as 97%. The new waist : 27" x 0.97 = 26.19", the new length: 30" x 0.97 = 29.1".

Janetta continued to look for bargains. She found cotton bed sheets at a great sale. She was interested in the "full size" fitted sheets. They were sized at 54" x 75". Again, there was a label inside that read "guaranteed 2% maximum shrinkage." Janetta knew that her mattress size measured 53" x 74". How much shrinkage should she expect on these sheets?

If you know that there will be 2% shrinkage, then, the actual remaining size can be found as the full size less the 2%: 100% - 2% = 98%.

Janetta calculated the new dimensions using this 98%, since the percent, P, is 98% and the base, B, will be the dimensions 54" x 75". The new sizes would become 54" x 98% = 54" x 0.98 = 52.92" and 75" x 98% = 75" x 0.98 = 73.5".

It appears that after shrinkage, the sheets will be too small! What size cotton bed sheets with maximum shrinkage of 2%, should she look for, to fit her bed? Or, another way of asking this question is: 98% of what amount (the size of the sheets) is equal to the mattress size?

The questions Janetta must ask are:
98% of what total size = 53"?
98% of what total size = 74"?

You are given the amount, A, in both cases, as well as the percent, P. We can solve for the two bases, B1 and B2. Here in this example, we set up the percent proportion, then cross multiply.

$$\frac{53}{B1} = \frac{98}{100} \qquad\qquad \frac{75}{B2} = \frac{98}{100}$$

Cross multiply, and solve for B1 and B2:

98B1 =	5300	98B2	=	7500
B1 =	5300/98	B2	=	7500/98
B1 =	54.08"	B2	=	76.53"

Janetta should buy her cotton bed sheets which have dimensions larger than 54.08" and 76.53". Now she can be a wiser shopper.

Type 7: Profit/Loss in Stock Market

The stock market is always in the news. It goes up, it goes down, and reporters announce percent gains and losses. Here are some examples of profits and losses in the stock market.

Example 11-19. This is a percent problem involving profit in the stock market. Josie bought a stock on the stock market for $60. After 3 months, she sold it for $90. What is her percent profit per annum? (That is, what is her percent profit for the year based on this transaction?) First, what is her profit for the three months?

Sales price - Cost price = Profit, or $90 - $60 = $30.

Her percent profit for the three months is found by using the original price as the base, B, and the amount of profit, $30 as the amount A. Use shortcut #2. Set up the ratio: A/B, 30/60 = 0.50 which is 50%.

But this transaction was for 3 months, one-quarter of a year. Consider this: 1/4 of a year's profit = 50%. Then for a full year's profit, multiply the 1/4 year's profit by 4:   4 · (1/4 year profit) = 4 · (50%) = 200% per year (annum).

Of course, one can lose money in the stock market. Here is Ann's story.

Example 11-20. This is a percent problem that involves loss on the stock market. Ann bought some stock on the stock market for $250. After 6 months, she sold it at a loss for $150. What was her percent loss per annum?

What is her loss for the six months (half of a year)? Sales price - Cost = Profit. However, when the cost is greater than the sales price, there is no profit. There is a loss. Ann has a loss in this business transaction! Losses in business are sometimes written in red ink, in parentheses, or with a minus sign in front of the number.

$$\$150 - \$250 = -\$100 \text{ or } (\$100).$$

This is a business loss of $100 for the six months. Her percent loss for the six months is found by using the original price as the base B and the amount of loss, $100, as the A. Use shortcut method #2. Form the fraction: A/B and change to a percent. This will give us the percent for the six months. We use the "minus sign" in these calculations to remind us that there is a loss. A/B = -100/250 = -0.40 = -40% loss for six months (1/2 of a year). P = -40%.

Multiply by 2 (2 halves in a year) to obtain the loss for the year: $2 \cdot (-40\%) = -80\%$ per annum. Bad luck, Ann!

The next example is a true story. A friend of mine asked me to figure his percent increase. This example is for you, Ken.

Example 11-21. A while back, Ken told me he had purchased Microsoft stock 2½ years earlier at a purchase price of about $3,400. He has not sold it, and he told me that it was now worth (approximately) $14,000. He wanted to know his overall percent gain and his average yearly percent gain.

His overall gain is found by subtracting the present value from the cost: $14,000 - 3,400 = $10,600, over a period of 2½ years.

The percent gain is found by using the original price as the B, and the amount of gain or profit, as the A. We solve for the percent, P, using shortcut method #2. Set A/B and change the decimal answer to a percent. You can use a hand calculator to find the value.

$$\$10,600/\$3,400 = 3.118 = 311.8\%$$

This percent gain is for the 2½ period, so that the annual percent gain is: 311.8% ÷ 2.5 = 124.7%

That's very good, Ken. Keep up the smart investing!

Business pages in newspapers list percent changes of the prices of stocks on the stock market. The percent change is the net change in price divided by the last quoted price, to one decimal place. If the stock price were to go down, there would be a negative change (-) and a negative percent (-%). If the stock price were to go up, there would be a positive change (+) and a positive percent (+%).

Example 11-22. Stock Market Percent Problem. The percent changes listed for certain stock market indices (namely the Dow Jones Industrials, the Dow Jones Transportation, and the Standard and Poors 500) were reported on a given day as shown in the following table.

Index	Last Price	Net Change	% Change
Industrials	7539.27	+60.77	
Transportation	2671.12	-27.88	
S & P 500	865.27	+2.36	

Find the % change for each of these indices. Remember, we will be approximating the answers to one decimal place. The amount, A, in all cases is the net change. The base, B, is the last price, as given in the above table. Find the different percentages, using shortcut #2.

For the Industrials, A = 60.77, B = 7539.27. Solve for the fraction: A/B (use your hand calculator) 60.77/7539.27 = 0.00806. Change this decimal to a percent: 0.00806 = 0.8% approximately.

Dow Jones Transportation: A = -27.88, B = 2671.12. In similar manner, you can find the value for the percent, P. A/B = -27.88/ 2671.12 = -0.0104375. Change this decimal to a percent: P = -1.04375% but written to one decimal place P =
-1.0%. Remember that this number is a negative change and must have a (-) sign: -1.0%.

Standard and Poors 500: A = +2.36, B = 865.27. We find P in the same manner as above to be 0.27% or approximately 0.3%.

Index	Last Price	Net Change	% Change
Industrials	7539.27	+60.77	0.8%
Transportation	2671.12	-27.88	-1.0%
S & P 500	865.27	+2.36	0.3%

On October 27, 1997, the U.S. Stock Market Dow Jones Average dropped 554.26 points from its relative high of 7715.41. Let's determine what the overall percent loss was.

Example 11-23. Stock Market Moves. On October 27, 1997 the actual loss was 554.26, and the new Dow Jones Index, after the drop was: 7715.41 - 554.26 = 7161.15.

To find the percent loss, we determine that the base, B, is the original Dow Jones Average (that is, 7715.41) and that the amount, A, is the drop of 554.26. Use shortcut #2, and find A/B: 554.26/7715.41 = 0.071838 a decimal number. Change this to a percent, P: 0.071838 = 7.18% ~ 7.2% percent drop.

This drop in the stock market Dow Jones Average was not the largest. Here is some data:

October 19, 1987	22.6% drop
October 28, 1929	12.8% drop
October 29, 1929	11.7% drop
October 27, 1997	7.2% drop.

(Do you think there is some relation to the Dow Jones Average "Witchings" and Halloween?)

On October 28, 1997, the very next day after the 7.2% drop, the New York Stock Market Dow Jones Average went up 337 points. Now, what was the percent increase over the previous day's closing price?

Remember, the stock market went down from 7715.41 to 7161.15, a drop of 554.26 in one day, a drop of 7.2% on October 27, 1997. On October 28, 1997, the stock market went up 337 points *from* 7161.15, and closed at 7161.15 + 337 ~ 7498.

To find the percent increase use shortcut #2 and find A/B where A is the amount of increase 337, and B is the previous days' Dow Average of 7161.15. A/B = 337/7161.15 ~ 4.7%.

Not a bad recovery!

Some stocks are very volatile, which means they have large changes, and often! Here is another stock market problem.

Example 11-24. Stock QQQ on the New York Stock Exchange was selling for $24 on Monday. On Tuesday, it went up 2 points (that is stock market talk for $2) to $26. On Wednesday, the stock fell 3 points, on Thursday it fell another 2 points. Then, on Friday, the stock went up 7 points.

In Part 2, when I discuss graphs, we will use this data to show you a bar graph and a line graph so that you will understand the data and get an instantaneous review of the stock QQQ for the week.

For this example, we would like to know the percent changes each day in the price of the stock, as well as the overall weekly change. We can set up a table as follows:

Day	Change (+ or -)	Price	Percent Change
Monday		$24	
Tuesday	+2	$26	
Wednesday	- 3	$23	
Thursday	- 2	$21	
Friday	+7	$28	

We fill in the percent changes as we find them. Then, we can find the overall percent change.

The amounts, A, are the changes. The base, B, are the previous day's prices. We start with Tuesday. Use shortcut method #2 to find the percents. Thus, divide A by B, multiply by 100. That will be P.

Percent change for Tuesday: (2/24) x 100 = 8.3%.
Percent change for Wednesday: (-3/26) x 100 = -11.5%.
Percent change for Thursday: (-2/23) x 100 = -8.7%.
Percent change for Friday: (7/21) x 100 = 33.3%.

The overall change for the week is 28 - 24 = 4 points. The overall percent change for the week: (4/24) x 100 = 16.7%. Of course, if you bought the stock on Thursday ($21) and sold it on Friday ($28) you did very well, indeed! Now we can complete the table.

Day	Change	Price	Percent Change
Monday		$24	
Tuesday	+ 2	$26	+8.3%
Wednesday	- 3	$23	-11.5%
Thursday	- 2	$21	-8.7%
Friday	+7	$28	+33.3%
Overall	+4		+16.7%

Type 8: Assorted Percent Problems

There are many percent problems that we encounter in our everyday lives.

Example 11-25. A restaurant gives senior citizens a 10% discount off their meals. However, there is still a 10% sales tax on the restaurant food.

Mr. Jones has dinner and his food bill total is $25. Let us find the amount of the discount, the amount of the sales tax, and just how much Mr. Jones has to pay for his dinner.

First, his discount of 10% is taken off his food bill. The base, B, is $25, the percent, P, is 10%. We have to find the new discounted amount, A.

Change the percent to a decimal, and multiply by the base: 10% = 0.10. Thus: 0.10 x $25 = $2.50 discount.

Mr. Jones' dinner bill is now: $25.00 - $2.50 = $22.50. You could use the one-step method here? If Mr. Jones received a 10% discount, then his bill would be: 100% - 10% = 90%, where 100% represents the whole bill. Then, change this percent to a decimal, and multiply by the base: 90% = 0.90. Thus 0.90 x $25 = $22.50.

You can work this shortcut #1 on your hand calculator. Key is the amount 25, press X, key in 90, press the % key if you have one. The answer: 22.50 will be in the display.

Now, let us work out the rest of Mr. Jones' bill. The sales tax of 10% is going to be figured on the net dinner bill, which came to $22.50. In this percent problem we have the percent, P, as 10%, and the base, B, as $22.50.

Use shortcut method #1 to find the amount of the sales tax. Change the percent to a decimal and multiply by the base. Thus: 10% = 0.10, and 0.10 x $22.50 = $2.25 sales tax. This tax is added to the discounted bill of $22.50, so that: $22.50 + $2.25 = $24.75.

Or, you can use the one-step method: 100% of the net plus 10% more of the net is the total bill. 100% + 10% = 110% = 1.10. Multiply this decimal by discounted price of $22.50. 1.10 x 22.50 = $24.75.
Well, Mr. Jones, at least you saved $0.25!

We hear a lot of weather forecasters giving us statistics about percentages of rainy days, snowy days and so on. Here is a percent problem involving the percent of rainy days in a certain month.

Example 11-26. There were 18 rainy days in the month of April. What percent of the days in April were rainy?

The number after the *of* is the base B, the total number of days in April, which we know is 30 days. The amount of rainy day is A = 18. What variable is missing? We are missing the percent P, (What percent of...).
Thus, we can use shortcut method #2 to find P. Divide A by B, and change the decimal to a percent. 18 ÷ 30 = 0.6 = 60%.
Thus, 60% of the days in April were rainy days.

Baseball is our national sport. People read the sports columns and chat about team standings. How are these team standings obtained?

Example 11-27. In baseball, the percentage listed in the sports pages (PCT) is the number of games won, divided by the *total number of games played*, calculated to three decimal places.

One spring day, the following teams had standings as follows:

Teams	Won	Lost	PCT
San Francisco	35	27	
Colorado	34	29	
Dodgers	30	32	
San Diego	27	34	

Find the PCT for each team.

For each team, the percent is unknown. A will be the amount of games won, and B will be the total number of games played.

San Francisco played 62 (35 + 27) games and won 35.

Colorado played 63 (34 + 29) games and won 34.

Find the PCT for each of these teams. Find A/B, change the decimal answer to a percent.

For San Francisco, A = 35, B = 62. A/B = 0.565.

For Colorado, A = 34, B = 63. A/B = 0.540.

In like manner, we find the PCT for the Dodgers to be .484 and for San Diego to be .443.

Now we can complete the table as follows:

Teams	Won	Lost	PCT
San Francisco	35	27	.565
Colorado	34	29	.540
Dodgers	30	32	.484
San Diego	27	34	.443

You read about opinion polls and surveys, and you note that the results are quoted as percents. Here is an opinion poll (survey) problem.

Example 11-28. A recent survey by an opinion poll company reported that they had contacted 85% of the residents in town. So far, 1,870 people had been called. What is the population of the town?

The total population is unknown, that is the base, B. The number of people contacted is the amount, A, which is 1,870. The percent is 85%. Set up the percent proportion:

$$\frac{1{,}870}{B} = \frac{85}{100}$$

Cross multiply, solve the percent proportion:

$$85B = 187,000$$

$$then \ \frac{85B}{85} = \frac{187,000}{85}$$

so that $B = 2,200$ *total population.*

Here is another survey problem. It deals with how many people are watching cable TV.

Example 11-29. This is a two-part problem involving a TV survey. In both parts, A is the unknown variable. In a recent survey of 1,500 people, 35% of them said they watched cable TV. How many watched cable TV?

Of those who watched the cable TV, 60% were men. How many men watched cable TV?

The total number of people, B, is 1,500. The percent, P, is 35. Find A, the number watching TV.

Use shortcut #1, change the percent to a decimal, then multiply by the base: 35% = 0.35, 0.35 x 1,500 = 525 people.

Now, use 525 as the total number of TV watchers, which is the new base B. Then, P = 60%, the percent of TV watchers who are men. We can find the number of men watching cable TV, which is the Amount, A, using the shortcut method. Multiply: 0.60 x 525 = 315 men.

Understanding and working these percent problems help us become more aware as consumers. The following example is based on true advertising.

Example 11-30. A certain automobile manufacturer issued its own credit card and announced that it would allow 5% of the amount charged to the credit card, to be applied on the purchase price of one of their cars. The lowest price car they manufacture costs $12,000.

Kevin ordered one of these credit cards and now wants to know how much he must charge on the credit card so that he will receive a car free (the lowest price car)! If Kevin charges $1,000 a year on the credit card, in how many years will he be eligible for a free car?

The first part of this example can be handled as follows: 5% of what number will be the price of the lowest priced car, $12,000? Thus, A is 12,000, P is 5%, and B is the unknown value. Set up the percent proportion and cross multiply to find B.

$$\frac{12,000}{B} = \frac{5}{100}$$

$$5B = 1,200,000$$

$$B = \$240,000$$

Kevin would have to charge $240,000 on his credit card to be eligible to receive a $12,000 car, free! Can you believe it?

If Kevin uses the credit card and charges $1,000 a year, he will be eligible for a free car in $240,000 ÷ $1,000 = 240 years! *What a bargain!*

Example 11-31. Here is a percent application involving advertising budgets.

About 52% of the total cost of a commercial product is spent on advertising. If the advertising cost amounts to $2,600, what is the total cost of the product? The total cost is unknown, B ("of the total cost"). The amount of advertising cost is A, and the percent is 52%. Set up the percent proportion:

$$\frac{2,600}{B} = \frac{52}{100}$$

Cross multiply to solve the percent proportion:

$$52B = 260,000$$

$$then \; \frac{52B}{52} = \frac{260,000}{52}$$

so that B = $5,000 *total cost of product.*

To check this answer, we read the problem as follows: 52% of the cost of the product is $2,600. Is 52% of $5,000 = $2,600? Use the shortcut method to check this: $0.52 \cdot 5,000 = 2,600$. The answer is correct.

Type 9: Income Taxes

It goes without saying that taxes play a large part in our everyday lives. In this section, I will work out some income tax problems. The first example, Example 11-32, will present situations for three single people. We will use the Federal Income Tax Table (for singles) Schedule X, and show how to calculate taxes due for these people, given their taxable income. Then, Example 11-33 will continue with the same data (from Example 11-32); however, you will use math skills developed in this book to find the *actual* percent tax on the same three peoples' incomes.

In addition, in Example 11-34, you will see how the "marriage tax" can affect you.

We will need the data and rates from the following Tax Rate Schedules for Federal Income Taxes. The data is correct for 1998.

Here is the Federal Income Tax Rate Schedule X for those persons filing their tax as singles. There are five different categories or income brackets. For most income brackets (for example, between $23,350 and $61,400) there will be two parts to the income tax due. First, there is a fixed amount for the tax due on the lower amount ($23,350). In this bracket, the fixed amount of tax is $3,802.50. In addition, there is a tax that is a percent of that income over the lower value ($23,350). In this bracket, the percent is 28%.

These tables are shown here as they appear in booklets sent out to the taxpayers. Notice, the heading:

Over	but not over	the tax is	of the amount over

It would be more understandable if we read the heading as:

more than	but not more than	the tax is	of the amount more than

Tax Rate Schedule for Singles

If taxable income is			
over	but not over	the tax is	of the amount over
$0	$25,350	15%	$0
$25,350	$61,400	$3,802.50 + 28%	$25,350
$61,400	$128,100	$13,896.50 + 31%	$61,400
128,100	$278,450	$34,573.50+ 36%	$128,100
$278,450		$88,699.50 + 39.6%	$278,450

Here is the Federal Income Tax Rate Schedule for married persons filing jointly. That is, the married persons combine their taxable income and taxable expenses. Then they use the following table.

Tax Rate Schedule for Married Filing Jointly

If taxable income is			
over	but not over	the tax is	of the amount over
$0	$42,350	15%	$0
$42,350	$102,300	$6,352.50 + 28%	$42,350
$102,300	$155,950	$23,138.50 + 31%	$102,300
$155,950	$278,450	$39,770 + 36%	$155,950
$278,450		$83,870 + 39.6%	$278,450

Now, let us work some of these tax examples.

Example 11-32. Tax Rate Schedules for Federal Income Taxes. We calculate the tax for several people and will refer back to the tax tables on the previous pages.

Dan has a taxable income of $23,250
Gayle has a taxable income of $48,700
Manuel has a taxable income of $130,000.

85

Dan's Tax: Dan's tax is found using the first line of the Tax Rate Schedule, as follows. His income is less than $25,350, therefore he pays 15% of his taxable income of $23,250. The amount (A) is the unknown value, B is $23,250, and P is 15%. Use the shortcut method to find A: 0.15 · 23,250 = $3487.50. Dan's income tax is $3487.50.

Gayle's Tax: Gayle's tax is found using the second line of the Tax Rate Schedule, since her income of $48,700 is more than $25,350 but not more than $61,400. She has to pay $3,802.50 + 28% of the amount over $25,350. Find the amount over $25,350. Gayle's taxable income is $48,700. The amount *more than* $25,350 is: $48,700 - $25,350 = $23,350.

Now, according to the instructions, Gayle owes $3,802.50 + 28% of $23,350.

We find 28% *of* $23,350. A is the unknown value, B is 23,350, and P is 28%. Use shortcut method #1 to find A: 0.28 · $23,350 = $6,538. Thus, Gayle owes $3,802.50 + $6,538 = $10,340.50.

Manuel's Tax: Manuel's tax is found using the fourth line of the Tax Rate Schedule. His income of $130,000 is more than $128,100, but not more than $278,450. He has to pay $34,573.50 + 36% of the amount over $128,100.

The amount *over* $128,100 is: $130,000 - $128,100 = $1,900. Now, according to the instructions, Manuel owes: $34,573.50 + 36% *of* $1,900.

We must find 36% *of* 1,900. Use shortcut method #1 where A is the unknown, B is $1,900, and P is 36%. Thus, 0.36 · $1,900 = $684. Manuel owes $34,573.50 + $684 = $35,257.50.

We can use the information from Example 11-32 to obtain additional information about the taxes our three friends have to pay.

Example 11-33. What percent, P, of their income must Dan, Gayle, and Manuel pay in income taxes as calculated from Example 32?

	Dan	Gayle	Manuel
Taxable Income	$23,250	$48,700	$130,000
Federal Tax Due	$3487.50	$10,340.50	$35,257.50

In each case, P is missing, and A and B are given. Use shortcut method #2, in which we calculate A/B, change the resulting decimal to a percent (move the decimal point two places to the right, and put in the % symbol). The answer will be the percent, (P).

Dan had an income of $23,250, and taxes of $3,487.50. A, the amount is $3,487.50, B, the base income, is $23,250. A/B = $3,487.50/ $23,250 = 0.15. Therefore, the percent, P, is 15%.

Gayle had an income of $48,700 and paid $10,340.50 in taxes. A/B = $48,700/$10,340.50 = 0.212 or 21.2%.

Manuel had an income of $130,000, and paid $35,257.50 in taxes. A/B = $130,000/$35,257.50 = 0.271 = 27.1%

Manuel can say that 27% of his (taxable) income is paid out in federal income taxes!

Here is the summary table showing what percent of their taxable incomes Dan, Gayle and Manuel paid in taxes. (Note that the symbol ~ means *approximately*.)

Summary Table

	Dan	Gayle	Manuel
Taxable Income	$23,250	$48,700	$130,000
Federal Tax Due	$3487.50	$10,340.50	$35,257.50
Percent of Taxable Income	15%	~21%	~27%

Example 11-34. Taxes for Singles and Married People. We will use the two tables for this example: the original Singles Rate Table for 1998 (see page 90) and the table of Tax Rates for Married Filing Jointly (see page 91). This example will illustrate what is meant by the "marriage tax." It really surprises many people!

Jack and Jill are engaged to be married. Jack has a taxable income of $51,000. Jill has a taxable income of $53,000. As singles, this is the income tax they have to pay (use the table for singles on page 84 line 2):

Jack calculates his income tax as $3,802.50 + 28% of ($51,000 - $25,350). So that his tax is: $3802.50 + 0.28($25,650) = $3802.50 + $7182 = $10,984.50.

Jill calculates her income tax as $3,802.50 + 28% ($53,000 - $25,350) so that her tax is: $3,802.50 + 0.28

% of ($27,650): $3,802.50 + $7,742 = $11,544.50.

Together, as singles, their total income tax is: $10,984.50 + $11,544.50 = $22,529.

Now, Jack and Jill marry! They have the same taxable earnings. Now, they calculate their income tax due as "Married, Filing Jointly," which is found in the second table on page 85. They use their combined income of $51,000 + $53,000 = $104,000.

Now they use line 3 of the table as follows: $23,138.50 + 31% of ($104,000 - $102,300) so that they have: $23,138.50 + 0.31% of ($1,700). Thus, $23,138.50 + $527 = $23,665.50.

Together, as a married couple, they pay $23,665.50. (When they were single, they had a combined tax of $22,529.) This is $1,136.50 more than they paid when they were single.

The next example is just one type of an example dealing with a person putting aside some earnings for his or her retirement. There are many calculations using percents when you start to plan for retirement.

Example 11-35. Every year, Marty can put aside 15% of his income in a special Retirement Account at his bank. How much must he earn to save $6,000 yearly in this account?

In this example, the P is 15%, the base, B, ("of his income") is what is unknown, and the amount, A, is $6,000.

Set up the percent proportion and cross multiply as follows:

$$\frac{6,000}{B} = \frac{15}{100}$$

$$15B = 600,000$$

$$B = \$40,000$$

Marty must earn \$40,000 yearly to save \$6,000 in his retirement bank account. We can check this amount. 15% of \$40,000 is found:

$$15\% = 0.15, \ 0.15 \cdot \$40,000 = \$6,000$$

Dear reader, we have completed Part I of the book. You have seen the solutions to a wide variety of percent problems. These included many types: commissions, interest, discounts, passing/failing, sales taxes, percent increase/decrease, profit/loss in the stock market, as well as income taxes.

In Part II, you will be introduced to Charts and Graphs, and will also do some line graphing. Some data will be presented in percent form, and sometimes you may have to calculate the percents to use in the graphs.

These skills that are presented in this book are used in everyday activities.

Part 2: Charts and Graphs

Chapter 12
An Introduction to Charts and Graphs

You know the old saying: "A picture is worth 1,000 words."

You can read pages and pages to explain a concept, yet one chart or graph can give you a lot of information all at once. Applications for charts or graphs can include: trends in budgets, sources of income, government spending, population changes, trends in stocks and bonds, results of test taking, home sales (average dollar per square foot), changes in wages and salaries (yearly), and many others. Of course, you have to understand how to read the graph or chart.

Perhaps one type of chart that you are familiar with is your city, state, or country map. It is a pictorial representation (a picture) *drawn to scale* of the area in which you live.

What is meant by *drawn to scale*?

Let's say there is a map of a city, that includes streets, avenues, parks, schools, etc. Or, you may have a map of your state that shows the interstate highways.

There is a scale marking on the map that may read: "1 inch represents 2 miles." Or, you may have a line with markings that would show what the equivalent size in inches on the map is for 1 mile, 5 miles, 10 miles, etc.

How did the map makers (cartographers) make the map? It was drawn using the math skills of ratios and proportions, just as we discussed in Chapter 8. The cartographers would say: If our scale drawing is 1 inch : 2 miles (1 inch on the map represents 2 miles in the city), then we can write this ratio as follows:

$$\frac{1 \text{ Inch}}{2 \text{ Miles}}$$

Now, if the main street (let's call it Broadway), is 5 miles long, how long should we draw the line on the map?

We can set up a proportion as we did in Chapter 9. This same concept of proportions works very well with a great many everyday applications. Thus we have:

$$\frac{1 \text{ Inch}}{2 \text{ Miles}} = \frac{X \text{ Inches}}{5 \text{ Miles}}$$

Cross multiply - that is, multiply the means by the extremes in this proportion. Then: $2X = 1 \cdot 5$, so that $X = 2.5$ inches.

In like manner, these cartographers work out their proportions for all the streets, avenues, etc.

Now, when you read the map, and see that Main Street measures ½" in length, you can determine the actual length of that street in miles.

$$\frac{1 \text{ Inch}}{2 \text{ Miles}} = \frac{1/2 \text{ Inches}}{X \text{ Miles}}$$

Cross multiply: $X = 1/2 \cdot 2$, $X = 1$ mile.

Charts and graphs are very useful in many other ways. I will discuss pie charts and bar graphs and line graphs. However, for all these graphs, we will have to know what type of data to compare. You will see that pie charts are useful when your data is in percents. You can chart your income and expenses; you can chart the portions that go for food, home, clothing, etc. as percents.

Real estate agents like to see charts of home sales over the past year, by months. They can see, at a glance, important information for their business. This data does not have to be represented as percents; real estate agents can use actual home sales figures. Bar graphs are very useful in this situation. Sometimes, real estate agents want to see home sales with additional information, such as by average price per square foot, yearly, or quarterly. Then, we can add another dimension to the bar graphs, as you will see later.

We can see trends, up or down, in certain stocks on the stock market, over a period of time.

We might want to set up a bar graph to show changes in our wages over a period of time. We can set up a single bar graph to show changes in wages of several people over a period of time.

The other chart we will discuss is the line graph. This consists of sets of points as input data. We connect these points with straight lines (hence the name "line graph"). Line graphs immediately show the trends, the up-swings and down-swings of such data as salaries, expenses, prices, sales, etc.

Pie Charts

Pie charts are fun to use and interpret. These graphs are pie shaped, and the *whole pie* is considered the whole amount of what we will be examining. For example, we discussed pie charts earlier in the book. Remember, we had pizza pies and we ate 4 out of 8 pieces of the pie. We could have eaten 3 out of 6 pieces of the same pie, because the whole pie is 1 or 8/8 or 6/6. Now that we also understand percents, we can say that the whole pie is 100%.

Pie charts are divided into various portions or slices, depending on the data; however, the pie chart's portions are represented as percents. No matter how we slice one pie, we cannot have more than 100%. This means that in order to set up a pie chart, we must make the portions or slices as percents of the total amount. We may have the original data in fractional form, or in text form. Such data must first be changed to percents.

Once you have the percent data, you can carve up the pie into those percent portions. For example, 25% of a pie would be 1/4 of the pie. Refer to Chapter 7, the tables of important percents, decimal, and fractional equivalents.

Here are a few examples of how a pie chart can be sliced up.

You can slice a pie chart in one-quarter, one quarter, and one half portions. You change these fractions to percents, so that the chart will show: 25%, 25%, and 50%. Here is that chart:

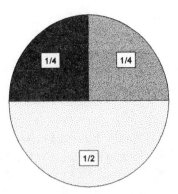

You can also slice the pie chart into 1/6, 1/6, 1/6, 1/4, 1/4 parts. Notice that the sum of these fractions: 1/6 + 1/6 + 1/6 + 1/4 + 1/4 = 1, the total pie. The equivalent percent for 1/6 is 16 2/3%, and the equivalent percent for 1/4 is 25%. If we add all the percents: 16 2/3% + 16 2/3% + 16 2/3% + 25% + 25% the sum is 100%, the whole pie! Here's how this pie would look.

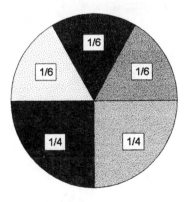

You see in the above pie chart that there are three slices of 1/6 each, and two slices of 1/4 each.

Let's try another example here. We'll just assume that the data is already in percent form.

Example 12-1. Mary has an earned income of $26,000. She has deductions of 8% for FICA, 10% for Income Tax, 7% for her IRA, 15% for her medical insurances. Her rent is $6,500/year which is 25% of her income. In addition, she has child care expenses of $2,860/year which is 11% of her income. Her food bills and utilities come to about $5,980/year, which is 23% of her income. The rest she tries to save for other emergencies.

We can tabulate Mary's expenses as follows:

FICA	8%
Income Tax	10%
IRA Savings	7%
Medical Insurance	15%
Rent	25%
Child Care	11%
Food, Utilities	23%
Total:	99%

So we see that 99% of Mary's income is allocated to her expenses. That leaves 1% of her income for miscellaneous spending.

Now, we can set up the following pie chart with these percentages.

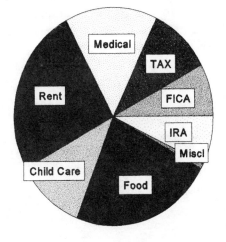

Bar Graphs

The bar graph can be a very informative type of chart. A bar graph consists of two (possibly 3, in three-dimensions) lines, mutually perpendicular to each other. These are called "axes"; there is a vertical axis and a horizontal axis. If the graph were a three dimensional one, there would be a third axis, mutually perpendicular to the other two. If you were to look at a corner of a room, where the walls and ceiling meet, you would have the three mutually perpendicular axes.

When creating a bar graph, you must know what we want to use as data. For example, you might want to see a chart of monthly car sales over a period of six months, or perhaps annual home sales in a certain

Whatever we decide, we must carefully label the axes with our meaningful data. Then we can draw in a bar (as shown in the graph in the next example), to represent the amount on one of the axes, extending from one axis and parallel to the other.

Example 12- 2. A car dealer has monthly car sales as follows:

Month	Cars Sold
January	150
February	200
March	175
April	250
May	100
June	100

We can set up a bar graph to show this information. Let the vertical axis represent the number of cars sold and the horizontal axis represent the months..

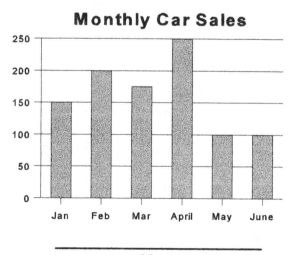

You can see why this type of chart is called a bar graph. The vertical lines drawn to meet the number of sales (along the horizontal axis) are drawn as bars.

The bar graph provides information at a glance. The car dealer can quickly tell how well his business is doing.

Example 12-3. Here is an another example of a bar graph in which we have some math test scores. Betty had the following scores on her tests: 85%, 90%, 80%, 75%, and 95%.

Here is the bar graph with the vertical axis as the test scores, and the horizontal axis as each individual test (Test 1, Test 2, etc.).

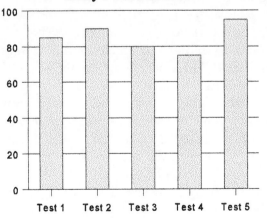

Bar graphs can be presented in various colors, and can even show a third set of data. Later in the examples in Chapter 13, we will present a three dimensional bar graph.

Interpreting Bar Graphs

Now, if you are given a bar graph, you may have to find the data and try to understand the trends that are shown.

Example 12-4. Consider a situation in which a company starts up in business. For the first four years, they lose money. This is can be shown as a negative number for their profits for those years. In algebra, negative numbers have a minus sign (-) in front of the number.

Nevertheless, the business hangs on. After a while, their product is accepted, and their profits start to increase. The following bar graph shows XYZ Company's profits over the past 10 years. As you interpret the graph, you can gain additional information about the company's performances. Notice that on this bar graph, the horizontal axis is in millions of dollars and the bars are horizontal.

Take a look at the XYZ Company Profits bar graph on page 101. What do you see? The XYZ Company had some pretty poor years, but now it seems to be in a very profitable situation.

You can read the values of profitability off the graph.

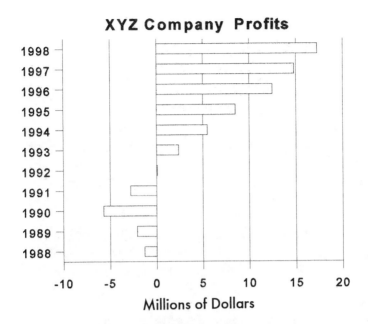

Look at XYZ's performance in 1997. The bar graph shows a profit at the 15 mark on the horizontal axis. Since this horizontal axis is in millions of dollars, the mark 15 refers to $15,000,000, or we can write the overall profit to be approximately $15 million.

The loss in 1991 was approximately $3 million. Investors in this company, or in the industry can draw their own conclusions

XYZ Company had losses in the years 1988 to 1991. There are several ways to show losses in business. Sometimes you write the amount of the loss, let's say a loss of $1,000, as: ($1,000). Sometimes, the loss is written in red ink! Algebraically, we put a minus sign (-) in front of the number. Thus, a $1,000 loss can be represented as -$1,000. (Note: Negative numbers are very important in algebra, and they are very useful in our everyday applications.)

Example 12-5. We sometimes can set up a bar graph and use data representing percents. Here you see a bar graph that has percents as its horizontal axis: 0%, 15%, 30%, and 45%.

The data in this graph is from the American Management Association, and it shows the percent of "information workers" using sources such as business or trade newspapers, personal contacts, World Wide Web, Electronic News, etc. to get the information they need for their job performances. The workers could use several sources, so they could be in two categories.

Notice that this is a three dimensional bar graph. The third dimension (that axis appearing to come out to the viewer) could be set up for different years over which the American Management Association obtains its data. We have data for one year.

Sources of Information

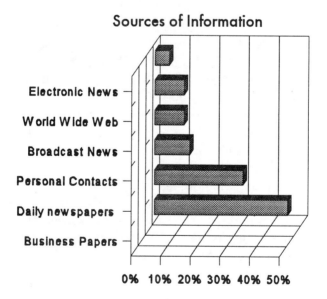

Line Graphs

You have learned about bar graphs and pie charts and that the bar graphs need vertical and horizontal axes. The user can decide what the axes represent.

You learned how pie charts are set up to use percentages of the whole pie. Now, let us consider line graphs. "Graphing" in mathematics usually refers to a form of line graph.

As with bar graphs, a line graph needs horizontal and vertical axes, which are labeled by the user. Now assume we have a rectangular system of two mutually perpendicular lines (2 dimensional). The horizontal line is called the "x-axis" and the vertical line is called the "y-axis". Where the x-axis and y-axis meet, or intersect, is called the "origin". We identify a point in this system as (x,y), an *ordered pair of numbers* and each such ordered pair uniquely defines a point on the graph. This rectangular coordinate system is called the *Cartesian Coordinate System* after its inventor, Rene Descartes.

Here is a method to take any ordered pair (x,y) and graph it on this rectangular system. This data could represent for example: (months, value of fund) or (years, cars sold), or in the following line graph, (number 1, number 2).

Now, instead of filling in bars, as we did for the bar graphs, we place a point, or a dot *at the intersection of where the data for the horizontal axis and the vertical axis intersects!* We place these dots wherever we have data points to put into the graph (the math people call these "plotting points").

After you have plotted the points in the graph, connect them together with straight lines. This set of lines is what forms our line graph. It will show trends and other information, as well as the bar graph.

Example 12-6. Here is a line graph showing the plotted points: (1,1), (2,3), (3,5) and the line connecting them.

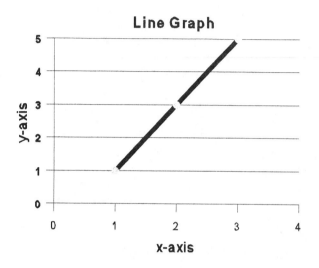

Example 12-7. Let us return to the car dealer who had monthly car sales for data that was used in a bar graph. Here is his sales data:

Month	Cars Sold
January	150
February	200
March	175
April	250
May	100
June	100

We can make a line graph as shown below, where the horizontal and vertical axes are the same as in the bar graph shown in Example 12-2.

This line graph shows the same information as the bar graph. However, the *trends* stand out. You see the peaking of the sales in April. You see the decline from April through June.

The slopes of the lines are very important, since they show the trends graphically. You can think of slopes as grades on hills as you bicycle along the roads. When the road is especially steep and difficult to pedal up, we say the road has a steep slope.

You can tell when you have a small slope, or even no slope at all. That is when you have a nice, flat road. Similarly, if you have a line graph with a horizontal line, that line has no slope (the slope is zero).

On the Monthly Car Sales line graph, notice the sales from March to April. The sales have increased, and the slope reflects the increase; it shows a line going up steeply. Then, after April, the sales decrease, and the slope of the line goes down steeply. But look at the line graph from May to June. It is a straight line. There is no slope. There was no change in sales for those two months as shown by the data.

This concept of slope is useful in mathematics, and works very well in these line graphs to show trends and activities.

Another way of thinking about slope is that slope is the ratio of the change in the vertical height to the change in the horizontal height.

Chapter 13
Pie Charts, Bar Graphs, and Line Graphs

In this chapter, we will work a variety of applications in pie charts, bar graphs and line graphs. There will be some applications that will have data applicable to *three* dimensional graphs.

Example 13-1. In one community, the real estate agent tabulated the number of homes sold for the year, as well as for the four previous years, by quarter, and created the following bar graph.

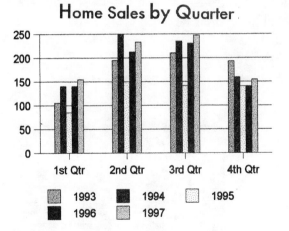

The agent had a lot of data in his computer, and this is what we see as a bar graph. Note that there are different shadings (it could be different colors, or different fill) for the five years tabulated. The vertical axis is the number of houses sold.

As you see, this bar graph contains a great deal of information, without using many words. That is the beauty of a graph or chart!

By looking at this graph, you can see several informative trends. For example, the number of homes sold was not particularly good in 1995. You can also see that sales improve in the second and third quarters as a general trend. This information may be important to a person who wishes to sell his or her house.

Example 13-2. The real estate agent has tabulated single family home sales by selling price over a three-year period. This time, the vertical axis is going to be labeled in hundreds of thousands of dollars. The horizontal axis is in quarters of the year.

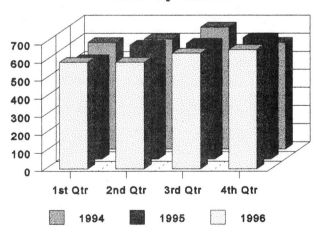

Sales by Quarter

The information given to us is represented on this three-dimensional bar graph. This graph gives us another way of looking at the data and its trends, the third axis shows the trends over the past three years trends.

Example 13-3. We have some world-wide data about sales of luxury cars. This bar graph represents sales, as the vertical axis, in thousands of cars sold. We also construct a line graph in which you may note the trends. For example, the numbers on this vertical axis, in thousands of cars sold become 3.1, 1.5, etc., representing 3,100, 1,500.

When we write 2.5, and the legend for the graph says "in thousands," we know we have to multiply 2.5 by thousands, (or 10^3). For the number 2.5, the graph data tells us that 2.5 x 1000 = 2,500 cars have been sold. (Move the decimal point in the number *3 places* to the right to multiply by 1,000.)

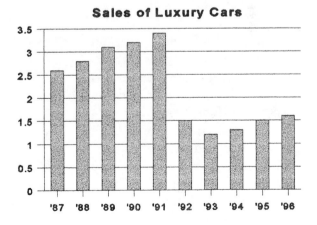

By showing the vertical axis as 1, 2, 3, etc. we can get more detail on our graph. The data is given as: 87: 2.6; 88: 2.8; 89:3.1; 90: 3.2; 91: 3.4; 92: 1.5; 93: 1.2; 94: 1.3;95: 1.5; 96: 1.6.

Here is a line graph using the same data we used in the bar graph

These luxury car models start out at £109,000 for the "reasonably priced" model, and can go as high as £230,000 for a luxury car with all the extras. They are British cars, and £ is pounds. How much would these models cost in U.S. dollars?

Let's assume that 1 £ is worth $1.50. Thus we can set up our ratios and proportions to find the price in U.S. dollars. So you see, even though we are in the section of the book dealing with charts, we are never far away from our ratios and proportions. Set up the proportions as follows:

$$\frac{\$1.50}{£1} = \frac{\$X}{£109,000} \qquad \frac{\$1.50}{£1} = \frac{\$Y}{£230,000}$$

These are two separate proportions. Perform the cross multiplication for each one to find the price of the two types of cars.

The least expensive of this luxury brand car is:

$$X = (1.50) \text{ x } 109,000 = \$163,500.$$

The highest priced of these cars is:

$$Y = (1.50) \text{ x } 230,000 = \$345,000.$$

At these prices, perhaps that is why the graphs show only thousands of cars sold per year! Notice that this example involved not only bar graphs and line graphs, we used ratios and proportions to convert the currency. We referred to "powers of ten" in order to label the vertical axis. We combine many concepts of mathematics to solve everyday, practical problems.

Example 13-4. A line graph sleight of hand. (It fools us.) Steepness of slopes is illustrated very well, using line graphs. Thus, it is even more important to observe the origin of the graph, so that we can better understand the trends illustrated. The origins of the graphs are where you start the x-axis (horizontal axis) values and the y-axis (vertical axis) at zero.

A Charity invested $97,000 in a mutual fund. The treasurer plotted the data of market value of the fund, over a period of six months, and showed the following line graph to his board of directors.

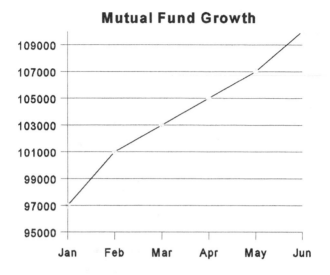

Mutual Fund Growth

Well, the board of directors were very impressed. Notice the steepness of the slope relating to the growth of the value the mutual fund. The board of directors were very pleased with their investment, because it increased from $97,000 to $110,000 in six months.

We can even find the percent gain for this fund. The percent gain for the six months is:

$$\frac{\$110,000 - \$97,000}{\$97,000} \cdot 100 = \frac{\$13,000}{\$97,000} \cdot 100$$

$$\frac{\$13,000}{\$97,000} \cdot 100 = 0.134 \cdot 100 = 13.4\%$$

The annual percent gain is 13.4% · 2 = 26.8%. That is a good yield on the investment.

Let's take another look at the same line graph. Now start the values on the y-axis with $0, as shown below.

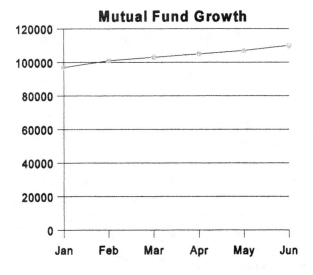

Mutual Fund Growth

Notice on this line graph, the slope of the data line is really not as steep as was initially shown to the board of directors.

We must be cautious with these data presentations, and carefully observe the axes at all times.

Dear reader, study these graphical examples. You will find data tabulated in graphical form in newspapers, and magazines. Graphs are used in businesses, as well as in fields of science, psychology, sociology, and so on. Reading and interpreting the results of a graph correctly is a very important skill in our everyday math and business lives.

Part 3: Using Algebra

Chapter 14
Introducing Variables and Constants

Many of our everyday problems can be solved using Algebra! Yes, dear reader, I've said the "A" word. Do not fear. But, let me give you a brief history of this remarkable branch of mathematics.

The name "algebra" comes from one of the works of Muhammed ibn Mūsā who wrote in Baghdad around the year 825 A.D. His work was titled: Al-jebr w'al-muqā bala. This meant: science of reduction and cancellation. We think this work referred to equations; and the name al-jebr was adopted by many later authors and changed into algebra.   But even before this work, there were treatises containing problems (which we now would call algebraic problems) that appeared in the Ahms Papyrus (which is in the British Museum). That dates from 1700-1600 B.C.!

In those days, they probably "solved" their equations by trial and error- try this number, try that number, see what works!

Algebra has come to be understood as representing numbers by letters, and using these numbers and letters in operations and relations. In a broader sense, algebra deals with equations and numerous other concepts.

Numbers: we are familiar with some of the mathematical number systems. We have already worked with: whole numbers, decimal numbers, fractions. The numbers we use represent quantities. Our numbers are fixed and do not change their value: the number 6 represents six items; the number 5.5 represents five-and-a-half items, and so on. In algebra, we call these numbers "constants."

Letters (Variables): Algebra uses letters of the alphabet-called "variables"-to represent numbers. Whereas the constants keep their fixed value, the variables are allowed to represent any particular number. Thus, they vary from problem to problem.

Operations: The mathematical operations we work with are addition, subtraction, multiplication, and division, respectively. The operator symbols we use are the symbols +, -, x, and ÷. These are called binary operators, and require *two* numbers, one directly before the symbol, and one directly behind the symbol. There are other mathematical operations, for example square roots, with its operator symbol √. This is a unary operation; it works on one number only.

Relations: We have mathematical symbols representing relations such as:

=	means "is equal to"
<	means "is less than"
>	means "is greater than"
~	means "approximately equal" (also ≈)

Algebraic Expressions: These expressions are combinations of constants and variables, using the algebraic operators of +, -, x and ÷.

Algebraic Equations: These equations are combinations of constants and variable, using the algebraic operators of addition, subtraction, multiplication, and division (+, -, x , ÷) as well as the relation"=" (equals).

Where do we use algebra in our everyday problems? We solve math problems using known algebraic formulas such as:

Rate x Time = Salary, (RT = S)
Rate x Time = Distance, (RT = D)
Area = Length x Width: (A = LW)

We can find averages, ratios, and solve some simple equations. We can work some statistical problems and probability problems that affect our daily lives. (Card players will see the need for certain probabilities.)

When we write formulas such as Area = Length x Width (A = L x W), we do not have to translate this from English into any other language. It is already in the universal language of algebra, and is recognized as an algebraic equation.

Assume the dimensions of your living room floor are given as: 12 feet by 14 feet. The dimensions of your kitchen floor is 11 feet by 12 feet. If you want to know the area of *each* room, you use the *same* formula, which is: Area = Length x Width, or A = L x W where A, L, and W are variables. The constants are the actual measurements of your rooms.

To find the area of your living room, use the constants for the living room as the L and W in the equation: 12 feet x 14 feet = 168 square feet. To find the area of your kitchen, use the constants for the kitchen as the L and W: 11 feet x 12 feet = 132 square feet. With this one formula, you can find the area of any rectangular surface (walls, floors, windows, etc.).

Chapter 15
Everyday Applications
in Algebra

We have discussed two of the many algebraic formulas that are used in everyday applications. Here are some other applications.

Example 15-1. Miles per Gallon. Suppose you are at a party and your friend Fred mentions that he gets 30 miles to the gallon on his new car! That is very good, but how does Fred get that number? To find the miles per gallon, here is what you have to do.

1. Fill up your car's gas tank with gasoline, write down the odometer reading (the number of miles driven). Let us call this mileage "M1."

2. Drive until the gas gauge reads 1/4 full. Then, fill up the gas tank again. This time, carefully record the number of gallons. Call the number of gallons G.

3. Write down the new odometer reading. Call this mileage "M2."

Now you can work the problem. First, find the miles driven between the two fill-ups.

The mileage is M2 - M1. The number of gallons of gasoline that were used is G.

$$\text{Miles per gallon} = (M2 - M1) \div G$$

Suppose it is March 15th, and you fill up the gas tank of our car and record the odometer reading at

84,021 miles. You drive the car for a week, and fill up the gas tank on March 22nd. The odometer reads 84,307. You have filled up the gas tank with 13 gallons of gas. What is the miles per gallon?

Miles per gallon = (M2 - M1)/G

Substitute your known values for M1, M2, and G (which were the variables in the equation)!

Miles per gallon = (84307 - 84021) ÷ 13
Miles per gallon = 286 ÷ 13 = 22 mpg

This mileage is not as great as Fred's, but at least you know that this mpg value is correct!

Example 15-2. In Example 15-1, you found that you were getting 22 m.p.g. driving your car. How much are you spending on gasoline per mile to drive your car? Let's find out:

a. How much you are spending per mile for gasoline.
b. How much you will spend on gasoline if you drive from Poughkeepsie, NY, to Richmond, VA, a distance of about 450 miles.

a. Let us assume that the price of gasoline is $1.76 per gallon. Set up a proportion as follows:

$$\frac{\$1.76}{22\ miles} = \frac{\$X}{1\ mile}$$

Cross multiply, so that 22X = $1.76, and X = $0.08.

If you use your hand calculator, enter 1.76, press ÷, enter 22, press =. You are spending $0.08 a mile for the gasoline.

b. To drive from Poughkeepsie to Richmond, you can use this formula to find the total cost of the gasoline. Total cost of gasoline = Cost per mile x Number of miles. Total cost = $0.08 x 450 = $36.00.

You can figure out other driving expenses. Use a one-year summary of a) number of miles driven and b) the total expenses (maintenance, tires, oil, registration, insurance, etc.). Total yearly cost ÷ Total miles driven = True cost per mile.

Example 15-3. Are there any poker players out there? There is a branch of mathematics (using algebra, or course) called probability and statistics that gives us information such as winning hands in poker.

We can understand some of the basic ideas and use some algebra to find a few statistics.

Poker is a game played with a deck of 52 cards, arranged in four suits (spades, hearts, diamonds, clubs) of 13 cards each. There are 13 face value cards: 2, 3, 4, 5, 6, 7, 8, 9, 10, Jack, Queen, King, Ace. Playing poker means selecting five cards out of the pack, and putting values on certain combinations of cards.

Some hands are much better than others. There are two hands that are interesting here. Consider a straight flush. This is the situation when you have one suit and five consecutive cards which are: 2, 3, 4, 5, 6; 3, 4, 5, 6, 7 all the way up to 10, Jack, Queen, King, Ace. (This last flush is called a royal flush because you have the royalty and ace.) We can discuss the probability of the various poker hands occurring.

If an event occurs, its probability is 1 (that is a sure thing). If an event cannot occur, its probability is 0. Probabilities, therefore, are decimal numbers between 0 and 1; the closer to 1, the higher the probability, or the more likely it is that the event will occur. Would you like to see the algebra associated with this concept?

If an event can occur in N (all possible ways), and if X of these ways are considered favorable (meaning they will yield the outcome we want), then the probability of the event occurring is given as: Probability of event occurring = X ÷ N.

Can we find the probability of an event *not* occurring?

If an event can occur in "N" (all possible) ways, in which "X" are favorable, then, using algebra, we can say that N - X are unfavorable outcomes.

To understand this algebraic notation, it is useful to substitute numbers for the variables, as follows:

Let the total number of ways an event can happen be 10 and let the number of favorable events be 7. (N = 10 and X = 7). Therefore, the number of unfavorable outcomes: N - X = 10 - 7 = 3.

Now we have this equation:

$$\text{Prob. of favorable event} = \frac{\text{No. of favorable outcomes}}{\text{No. of all possible outcomes}}$$

From our numerical example above, the probability of favorable event = 7/10; the probability of unfavorable event = 3/10. The sum of these two probabilities = 1.

That is: 7/10 + 3/10 = 10/10 =1. The sum of probability
of an event happening and probability of the same
event *not* happening is always equal to 1.

Notice how we used the algebraic concepts of
variables and constants, and the concept of fractions.

Now we will use decimal numbers. Let us get back
to the poker hands with flushes. Consider the straight
flushes. Notice that there are nine such possible
combinations. Also, there are four possible suits.
Therefore, you find that the total possible flushes are:

9 possible combinations x 4 suits = 36 possible events

Using statistical math, one finds that there are
2,598,960 possible poker hands. Honest! (See Appendix
2 for further details.)

What is the probability of getting a flush? The closer
the probability is to 1, the closer you are to getting the
event! Probability of a flush = total possible flushes/
total possible hands. = 36/2,598,960 = 0.00001385169.

This can be written as an approximate fraction:

$$\frac{1}{100,000}$$

which is read as 1 chance in one hundred thousand to
get a flush.

There are only 4 possible royal flushes (the four
suits), so that:

The probability of royal flush is = 4/2,598,960 =
0.00000153908. This can be written as an approximate
fraction:

$$\frac{2}{1,000,000}$$

which is read as 2 chances in 1 million to get a royal flush. And that's why there is such a big payout for flushes!

For more information on "4 of a kind," "Full House," "One Pair," and how to find the total number of all possible poker hands, see Appendix 2.

Example 15- 4. Working with Averages. When we use the word "average" in mathematics, we mean the following formula:

$$\text{Average} = \frac{\sum \text{ of Quantities}}{\text{Number of Quantities}}$$

where $\sum$ means "sum." In other words, the average of two numbers is:

$$\text{Average} = \frac{\sum \text{ of Two Numbers}}{2}$$

The average of three numbers is:

$$\text{Average} = \frac{\sum \text{ of Three Numbers}}{3}$$

Referring back to Betty's test scores on page 99, we had set up a bar graph for the tests: 85%, 90%, 80%, 75%, 95%. What is her average grade?

$$\text{Average Grade} = \frac{\Sigma(85\% + 90\% + 80\% + 75\% + 95\%)}{5}$$

$$= \frac{425\%}{5}$$

$$= 85\%$$

In another example, we also had a bar graph for a car dealer who had monthly sales that were shown on the graph as the following car sales:

Month	Cars Sold
January	150
February	200
March	175
April	250
May	100
June	100

What was the monthly average number of cars sold?

$$\text{Monthly Sales Average} = \frac{\Sigma(150 + 200 + 175 + 250 + 100 + 100}{6}$$

$$= \frac{975}{6}$$

$$= 162.5$$

Example 15-5. Here is a challenging problem on pricing groceries. The supermarket is advertising a special price of $2.70 for a brand name 18 ounce (oz.) box of cereal. They pack 24 boxes to the case.
a. How much does 1 ounce of this cereal cost?
b. How many pounds does a case weigh?
c. How much does a case cost?
d. This supermarket has its own special brand of cereal (very similar to the advertised brand) selling for $3.92 for a box that weighs 1 pound, 12 ounces. Is this a better buy than the advertised special?

a. You know that 16 ounces (oz.) = 1 pound (lb.). To find the price per ounce, divide: price/oz. which is $2.70/18 oz. Use your hand calculator to find that the price per ounce is $0.15.
b. There are 24 boxes of this special to a case, each box weighing 18 oz. Thus the weight of the case is: 24 x 18 oz. = 432 oz. To find the weight in pounds, divide 432 by 16. A case weighs 432 ÷ 16 = 27 pounds.
c. A case costs 24 x $2.70 (price per each box) which is: price per item x number of items = cost = $64.80.
d. Is this advertised special really a good deal? Let's find the unit price of the supermarket's own brand. The total number of ounces in this box is 16 + 12 = 28 ounces; it costs $3.92. Now, find: price/oz. which is $3.92/28 = $0.14. If you like the taste of this special brand of cereal, you can save money by buying it; it has the lower unit price.

Dear numerically challenged reader, algebra is logical, and once you understand the basic concepts, by proceeding in a stepwise manner, you will be able to work many different types of algebraic problems, and then be able to solve a variety of everyday applications.

Appendix I
Glossary of Math Terms

Algebra: That branch of mathematics dealing with equations, variables, and constants.

Algebraic Equation: Algebraic expressions connected by the equal sign (=).

Algebraic Expression: Those expressions that are combinations of algebraic constants and variables, and using algebraic operators (+,-, x, ÷).

Algebraic Formulas: Some algebraic formulas are algebraic representations of physical occurrences, using variables. We use formulas when we substitute our own constants for the variables to obtain meaningful answers.

Average:

$$\text{Average} = \frac{\sum \text{ of Quantities}}{\text{Number of Quanties}}$$

where $\sum$ means summation.

Axes: Two (or more) lines mutually perpendicular to each other. Used in bar graphs, line graphs, and mathematical graphing. In two dimensional graphs, there is a vertical axis and a horizontal axis. Where they intersect (usually at zero) the point of intersection is called the origin.

Bar Graphs: See Graphs, Bar Graphs.

Base, Base 10: In the number 10^2, the number 10 is called the base. The 2 is the exponent, written above and to the right of the base. This number 10^2 is the same as $10 \cdot 10 = 100$.

Cartesian Coordinate System: Developed by Descartes, this is a system of mutually perpendicular, vertical and horizontal axes, together with plotted points of the form (x,y).

Constant: Quantities that are fixed and do not change in value. For example, our number system where 5 represents five and only five.

Cross Multiply: To solve a proportion, use the property that the product of the means is equal to the product of the extremes. Some say that we *cross multiply*, because that is what is happening in the proportion.

$$\frac{A}{B} = \frac{C}{D} \text{, so that: } A \cdot D = B \cdot C$$

Charts and Graphs: Pictorial representation of data.

Decimal: A decimal number is a fractional part of a whole number whose denominator is a power of 10. The number of zeros in the denominator is always the same as the number of decimal places in the number. All whole numbers are decimal numbers with an understood decimal point to the extreme right of the number.

Decimal, Rounding and Truncating: Techniques that allow for satisfactory accuracy, yet shorten the resulting decimal places in the final answer by either cutting digits beyond a certain point or rounding off at a certain place value.

Denominator: When we have a certain object (say, a pizza pie), if we cut it into 8 *equal* pieces, that *denominates* or tells us into how many equal pieces we have cut the pizza pie. Therefore, that *denominator* is equal to 8. In a fraction A/B, the value B is the denominator.

Digits: Digits are any one of the ten Arabic numeral symbols: 0,1,2,3,4,5,6,7,8,9.

Drawn to Scale (also: Scale Drawing)**:** Ratio of a dimension on a map or chart to the actual dimension, for example, 1 inch = 2 miles.

Equation: A mathematical statement expressing the equality of two quantities. An equation has an *equal* sign (=). The left side of the equation "equals" the right side of the equation.

Equivalent Fraction: See Fraction, Equivalent.

Exponent: A quantity that tells how many equal factors there are, and is also called the *power*. The exponent is always written a little above and to the right of the base, for example 10^3, in which 3 is the exponent.

Extremes: See Proportion, Extremes, Means.

Factor: Factors are numbers that multiply each other.

Fraction: A Fraction is the ratio or comparison of two numbers, written as A/B, where the value of B is *never* zero. The value of A is called the numerator, and the value of B is the denominator.

Fraction, Equivalent: Fractions are equivalent when they represent the same amount. For example, 3/6 and 4/8 are equivalent fractions.

Fraction, Improper: fractions that have numerators *larger than or equal* to their denominator. For example 7/4, 9/5, 6/6, etc.

Fraction, Proper: Fractions that have numerators *smaller* than their denominator. For example 1/2, 3/4, 5/8, etc.

Graph: A diagram or picture representing any sort of relationship between two or more items by means of dots, lines, or bars. Graphs may be two or three dimensional, consisting of mutually perpendicular lines called axes, on which data is plotted.

Graphs, Bar: A two or three dimensional graph in which the data is entered by drawing a "bar" to represent the data quantity, extending from one axis and parallel to the other.

Graphs, Line: A two or three dimensional graph in which the data is entered by placing a "point" or dot at the intersection of where the data for the horizontal and vertical (and possibly a third) axes intersect. These points are connected by straight lines..

Horizontal Axis: The horizontal line that forms part of the mutually perpendicular lines that make up the *axes* in a graph.

Improper Fraction: See Fraction, Improper.

Line Graphs: See Graphs, Line.

Means: See Proportion, Extremes, Means.

Metric Conversion: Finding the equivalent measurements from British Units to Metric, and Metric to British Units, using ratios.

Metric System: A System of measurements (weights, lengths, volume, temperature) based on the decimal system. This is a very widely used system of measurements. The Unites States uses the British units of pounds, miles, gallons.

Mixed Number: A number that contains a whole number and a fractional part.

Numerator: When we *number* how many pieces we consider, out of the whole amount (for example we eat 3 pieces of pizza out of the whole pie, which is 8 pieces), the *numerator* is 3. In a fraction A/B, the numerator is the value of A.

Origin: The zero point where the horizontal and vertical axes intersect.

Percent: Means *parts per hundred*, and is represented by the symbol %.

Percent, Proportion: A/B = P/100, where A is the amount, B is the base, P is the percent.

Pie Charts: Chart in which a circle (pie shaped) is divided into portions, usually percentages, according to given data. The sum of the portions is a whole pie, or 100%.

Probability: Probability can be considered the fraction:

$$\frac{\text{number of successful results}}{\text{number of all possible results}}$$

Probabilities are decimal numbers ranging from 0 to 1, with 0 representing no successful results, and 1 representing all successful results. The larger the decimal, the larger the probability the event will be successful.

Proper Fraction: See Fraction, Proper.

Proportion: A statement of equality between two equal ratios. A proportion may be written as A/B = C/D, *where B and D are not zero*. Proportions are another way of writing two equivalent fractions.

Proportion, Extremes, Means: In the proportion A/B = C/D, the values of A and D are called the *extremes* of the proportion, because they are at the extreme positions of the proportion. The values of B and C are called the *means* of the proportion because they are in the middle positions of the proportion.

Ratio: The comparison of two numbers. The ratio of A to B can be written as A:B or A/B or A÷B, where B can never be zero.

Relational Operators: Give a relation between two values, such as one value, A, is equal to, less than, or greater than another value, B. Some relational operators are: " =, is equal to", "<, is less than", ">, is greater than", "~, is approximately equal to."

Scale Drawing: See Drawn to Scale.

Slope of a Line: The ratio of change in vertical height to the change in the horizontal length.

Variable: A letter or symbol representing a number or an unknown quantity.

Vertical Axis: The vertical line that forms part of the mutually perpendicular lines that make up the *axes*.

Whole Number: The natural counting numbers (1, 2, 3, etc.) as well as the number 0.

Appendix II
Optional Examples

Here are some optional examples as I promised you.

Example A-1. This optional example is about comparing prices of similar items when the items come in various sizes. For example, detergents and cereals come in various size boxes and containers of pounds, ounces, and fractional ounces for dry weight; and gallons, ounces, and fractional ounces for liquid measure. Detergents and cereals often come in medium, large, super, giant, and economy sizes. Do we really save money by buying the giant size?

We would like to make the best purchase and can make that decision if we knew the "unit price" of each item. The "unit price" is the price per pound or per ounce. We find the price per ounce for each item (*such as various sizes of the same cereal*), and compare the unit prices.

Thus, if your favorite cereal comes in various size containers, you will know which size will be the most economical. It may not necessarily be the giant size!

In measuring dry weight and liquid measure, recall that for:

Dry weight	16 ounces is one pound
Liquid measure	32 ounces is a quart
	128 ounces is a gallon

(If we were in the Metric System, we would not have such conversion problems! But we are still using the British System of Measure.)

Let us set up a table with the grocery items we buy and use the most. (See table on page 134.) I have gone to my local supermarket and have found the following weights and prices for cereals and detergents. I will call them Brands X, Y, Z, etc.

Fill in the columns with the name, the weight as it appears on the container, and the price charged by the store. In the fourth column, *change all the weights to ounces!* If the item contains fractional ounces, change those fractions to decimals.

If an item is listed at "3 pounds, 5 ounces," change the three pounds to ounces: 1 pound = 16 ounces; therefore 3 pounds is 3 x 16 = 48 ounces. The item weighs 48 + 5 = 53 ounces.

If an item is listed at 2 pounds, 4½ ounces, change the 2 pounds to 2 x 16 = 32 ounces. Change 4½ ounces to 4.5 ounces. The item weighs: 32 + 4.5 = 36.5 ounces.

If you have liquid measure, and have *1 gallon 2 quarts*, change the gallon and quarts to ounces.

There are 128 ounces in a gallon; 32 ounces in a quart. Thus, 2 quarts = 2 x 32 ounces = 64 ounces. Four quarts = 1 gallon = 128 ounces. The item, 1 gallon 2 quarts, is converted to ounces and is: 128 + 64 ounces = 192 ounces.

Now that we have standardized all the weights or volumes, we take the price (from the supermarket shelf) that has been entered in the third column. *Divide this number by the weight or volume in ounces.* Use your hand calculator for this, since you may have a number of calculations to do.

Enter this number (in the display of the calculator) into the fifth column of the table to *three decimal places.* This is the unit price! Compare unit prices of the same product, then make your educated choice!

Here, in this table I have listed prices for various sizes of Brand X, Brand Y, and Brand Z. On page 135 I have included a blank table for you to use for your own items.

Brand	Size	Price	Weight (Oz.)	Price/Oz.
Cereals				
Brand X	1 lb. 8 oz.	$3.79	24 oz.	0.158
Brand X	1 lb. 2 oz.	$2.85	18 oz.	0.158
Brand X	12 oz.	$2.15	12 oz.	0.179
Brand Y	1 lb. 2 oz.	$2.39	18 oz.	0.133
Brand Y	1 lb. 8 oz.	$3.15	24 oz.	0.131
Detergents				
Brand Z	50 oz.	$4.59	50 oz.	0.092
Brand Z	2 qts.	$6.85	64 oz.	0.107
Brand Z	3.12 qts.	$8.39	100 oz.	0.084
Brand Z	1.56 gal.	$13.99	200 oz.	0.070

Your Turn

Brand	Size	Price	Weight (Oz.)	Price/Oz.

Here is another optional example for you:

Example A-2. How do you calculate the price of yard goods (fabric, carpeting, drapery materials, etc.) when such items are sold by the yard (36 inches or 3 feet to the yard), and fractions of a yard, and the price is in dollars and cents (decimal numbers)?

Leesa wanted to get new carpeting for her living room. She measured the room and found it to be 15 feet wide and 18 feet long. She found nice carpeting for $25 per square yard. There was an additional $2.50 per yard for the carpet pad, and installation costs were $1.50 per square yard. How much will the complete job cost, if there is an additional 8% sales tax on the carpet and carpet pad?

Leesa's living room measures 15' by 18', or, if we change feet to yards (there are three feet in one yard), 5 yards by 6 yards. Now we know that she needs 5 x 6 = 30 square yards of carpeting and padding for her living room.

The carpeting will cost $25 x 30 = $750 and the padding will cost $2.50 x 30 = $75. There is an 8% tax on this total of $750 + $75 = $825.

The base, B, is $825, the percent, P, is 8%. Find the amount, A, using the shortcut method: Change the percent to a decimal and multiply by the base: 8% = 0.08; 0.08 x 825 = $66 tax.

Leesa also has an additional charge of $1.50 per square yard for installation. Thus, that charge is $1.50 x 30 = $45. The total amount for this job is: $750 + $75 + $66 + $45 = $936.

But wait a minute, perhaps Leesa could tile the floor for less money. She found a good tile company where she could get special tiles that were 1 foot squares each at a cost of $3 per tile. There is still that

8% tax on the materials, and there is a $250 installation charge. Now, how much would the tile job cost Leesa?

Her living room dimensions are 15' by 18'. Now we have to calculate the square footage! The area of her living room in square feet is 15 x 18 = 270 square feet. Leesa will need 270 tiles, because they are one-foot squares. The price of each tile is $3, so that 270 tiles cost $3 x 270 = $810.

The tax of 8% is found, using the shortcut method. Change the percent to a decimal, and multiply by the base which is $810. 8% = 0.08, 0.08 x 810 = $64.80 tax.

This tile job costs $810 (tile) + $64.80 (tax) + $250 (installation) = $1,124.80. Well, Leesa, you may have to do some more comparative shopping. As it turns out, the carpeting job would cost $936, and the tile job would cost $1,124.80.

The following example, A-3 will be a practical example of working with decimal numbers, namely dollars and cents. But first, let's review the steps in balancing a checkbook.

Your bank just sent you your monthly statement, which shows deposits that were made electronically, by mail, ATM, or in person. The statement shows the interest credited to your account as well as the withdrawals by check, electronically, by ATM, or in person. The bank may send you the canceled checks you wrote and were paid in this period. You will receive a full itemization of all your issued checks that were paid by the bank in this period.

The first thing to do is *not to throw up your hands and put the whole mess away for later!* I am going to walk you through a simple method so that you can "reconcile" your statement. This means that you will compare your records with the bank's statement and you make sure that they show the same ending balance. You will need to keep track of your deposits (all types), checks written, and other electronic withdrawals in a check register. This could be in the form of attached check stubs, lines in a small insert book, or whatever. Whatever type of register you use, when you write a check *you must record the date, the check number, the amount paid, and to whom paid!* In addition, you should subtract that amount from your check balance. In this way, you keep up-to-date in the register.

You need a starting point. Consult last month's bank reconciliation statement. You will also need a clean sheet of paper on which to record some of the data needed for the calculations. Then take the following steps.

1. Draw a line in your check register just after the last check that was listed by the bank, from your previous month's statement. This is your starting point.

2. Draw a line in your check register just after the last check written that appears in this month's bank statement. *This is your tentative balance.* Record it on the separate paper, and label it "tentative balance."

3. Consult this month's bank statement. This statement has all the information you need. Note all deposits (electronic, ATM, interest, checks, cash, etc.) that appear on your bank statement, then check them off (√) in your checkbook register one by one.

4. Note all withdrawals (electronic, ATM, cash, utility bills, etc.) and check them off (√) your check register.

5. Record on the separate paper what the bank statement shows as your current balance. Label this amount your "Current Balance."

Bank reconciliation statements will list all the checks that were paid out, in numerical order. These statements also show any break in the sequencing of the checks by putting an asterisk (*) if there is a break in the numbering sequence. This makes it easy for the user to see at a glance what checks did not get processed in the given month.

6. Now, go through the register, from the line drawn in Step 1 (above) to the line drawn in Step 2 and write down on the separate paper the amounts and check numbers of the checks you wrote that *do not appear* on this current statement. (These are the checks that have not as yet been cashed.) You know you have put money aside for these amounts, but they are not to be counted in for this statement. These are called "outstanding checks." Add them and record the sum!

7. There may be deposits you made *that occur after you drew the line for your tentative balance, and that appear in the bank reconciliation statement* (Step 3). Add only those deposits that occur *after* you drew the line, no others. Record the sum on the separate paper and call this sum "additional deposits."

Now you are ready to do the math. Add: tentative balance (Step 2) + additional deposits made (Step 7) + sum of outstanding checks (Step 6). This sum should be the same amount as the "Current Balance" as shown on the bank statement (Step 5).

What if it is not the same? You get your hand calculator and start to do the following: Starting with the line drawn after the last check you wrote in the previous month, you will have to do the addition and subtraction (addition for the deposits, subtraction for the withdrawals) for each entry. The bank statements are usually correct. It is the user who may have written a check or two without recording the amount in the register, may not have listed some withdrawals, or made a goof in arithmetic. Such an error can be found as you go from line to line and check the arithmetic.

Example A-3. Jon has a checking account. His bank sent him the monthly statement. His ending balance is $3,500 and the last check shown paid is check #642.

Jon opens his check register and draws a line under the transaction of check #642. He reads his *tentative balance* from his checkbook as $2,400.

Jon has checked the statement and notes that checks #603, #621, and #633 are not listed (outstanding checks). Jon looks up the amounts for the checks and writes them as: #603, $450; #621, $50; #633, $60. The sum is: $560.

Jon also notes that he made a deposit of $540 *after* check #642 that appears on the statement.

Jon reconciles his checking account. Tentative Balance + Outstanding Checks + Deposits after the last check = $2,400 + $560 + $540 = $3500. Correct!!

To complete this exercise, Jon makes sure he checks off all the listed withdrawals and all the listed deposits as outlined in Steps 4 and 5. Then, he will be ready for next month's reconciliation statement.

As discussed in Part 3 of this book, here are some more probabilities for other Poker hands. In the matter of probabilities of poker hands, remember that there are 52 cards in a deck, and a poker hand contains 5 cards. We would first like to find out the number of *different* ways five-card hands can be drawn from an ordinary deck of cards.

Consider 5 cards dealt, and call them: A, B, C, D, E. You certainly can have any one of the 52 cards for the A position. You can have any one of the remaining 51 cards for the B position, any one of the remaining 50 cards for the C position, any one of the remaining 49 cards for the D position, and any one of the remaining 48 cards for the E position!

The total number of ways of getting all possible hands is to multiply: $52 \cdot 51 \cdot 50 \cdot 49 \cdot 48$.

However, in the matter of possible different poker hands, for example, we could have the following hand: (where S = Spade, H = Heart, D = Diamond, C = Club): 5S, 7D, 6H, 4C, QD (that is, 5 of spades, 7 of diamonds, and so on).

But wait! We also could have:

$$7D, 6H, 4C, QD, 5S$$
$$6H, 4C, QD, 5S, 7D$$
$$4C, QD, 5S, 7D, 6H$$

as well as many other possibilities, all with the same five cards. The cards are the same; *the order is different.* In poker, the order does not matter. The hand 5S, 7D, 6H, 4C, QD is the same hand as the other three.

Thus, the product $52 \cdot 51 \cdot 50 \cdot 49 \cdot 48$ includes all the duplicated orders. In order to get a meaningful number (total of all possible poker hands, without order counting) we must modify the product. We can write the following product:

$$\frac{52 \cdot 51 \cdot 50 \cdot 49 \cdot 48}{1 \cdot 2 \cdot 3 \cdot 4 \cdot 5}$$

The product in the denominator has to be explained. If we only had one card to hold, then there would be:

$$\frac{52}{1} = 52 \; possibilities$$

If we had only 2 cards to hold, then there would be:

$$\frac{52 \cdot 51}{1 \cdot 2} = 1326 \; possibilities$$

The denominator of 2 is used to modify the product 52 • 51 so that the order does not count. For example, if you had: 7D and 5H, as first and second cards, it is the same as 5H and 7D as first and second cards. These two arrangements are the same. For two cards, then, this duplicating effect can be corrected by dividing by 2.

If we had only three cards to hold, there would be:

$$\frac{52 \cdot 51 \cdot 50}{1 \cdot 2 \cdot 3} = 22,100 \; possibilities$$

With three cards, they can be arranged in six different possible positions. Consider 5H, 7D, QS. We can arrange these three cards in only six possible ways:

5H, 7D, QS   5H, QS, 7D   7D, 5H, QS
7D, QS, 5H   QS, 7D, 5H   QS, 5H, 7D

In poker, all these six possible arrangements are the same, so that we use $2 \cdot 3 = 6$ as the divisor to eliminate such duplication.

Similarly, for a four-card hand, there are 24 ways for duplication, and therefore for the poker hand of five cards, there are $1 \cdot 2 \cdot 3 \cdot 4 \cdot 5$ duplications.

For our poker hand, therefore, we have:

$$\frac{52 \cdot 51 \cdot 50 \cdot 49 \cdot 48}{1 \cdot 2 \cdot 3 \cdot 4 \cdot 5} = 2{,}598{,}960$$

This is called a *"combination"* in probability, and is written as $_{52}C_5$ which is probability shorthand for: "choose 5 out of 52, without considering order."

Now, let us find the probability of a "full house."

This is an arrangement of the 5 cards as follows: 3 of one kind, 2 of one kind. Consider a full house to be: XXXYY. Remember that the suits are spades, hearts, diamonds, or clubs (S, H, D, or C).

For the XXX, which refers to the same face card, we can have S, H, D; S, H, C; H, D, C; S, D, C. You see there are 4 possible orderings, and there are 13 face cards available.

For the YY, which refers to a different face card from the XXX, we can have: S, H; S, D; S, C; H, D; H, C; C, D.

There are 6 possible orderings. Now, there are only 12 available face cards (Because we used one face card for the XXX.) Thus, we have $4 \cdot 13 \cdot 6 \cdot 12 = 3{,}744$ possible ways to get a full house. You notice that we multiplied all possible ways of getting the full house. Order counts in this instance.

The probability of getting a full house is the number of ways of success divided by the total number of poker hands: 3744 /2,598,960 = 0.00144, or 144/100,000 or approximately 1/1,000.

What about "4 of a kind"? Can we use some math to determine that probability? "Four of a kind" means you have a hand with 4 of one value, and one other card. That other card could be one of the remaining 12 values in any of the 4 suits. That one card, therefore, could be any one of the 48 remaining cards.

Your "4 of a kind" could be 4 of the 2s, 3s, 4s, 5s, 6s, 7s, 8s, 9s, 10s, Jacks, Queens, Kings, or Aces.

As you see, there are 13 possible sets of 4. With each set of 4, you could have any one of the remaining 48 cards as the fifth card.

The total possible ways of getting "4 of a kind" would be: 13 possible sets x 48 possible remaining 5th card = 624. Therefore the probability of "4 of a kind" is equal to (total "4 of a kind")/(total possible hands): 624/2,598,960 = 0.00240096038.

Compare the probabilities we have found:

Probability of Flush = 0.00001385169 (one chance out of 100,000).
Probability of Royal Flush = 0.00000153908 (two chances out of 1,000,000).
Probability of 4 of a Kind = 0.00240096038 (two chances out of 1,000).

Dear reader, these are decimal numbers, and as you see, the probability of "4 of a kind" occurring is a more likely event than that of the flushes, because the "4 of a kind" probability is a larger decimal number, and the closer to 1, the greater the certainty of occurrence.

Let's find the probability of getting one pair in poker. This configuration is: X X A B C where A cannot be an X, nor can B be an A or an X, nor can C be an X or an A or a B (because there is just one pair). The total number of ways of getting one pair is the product:

$$\frac{(6 \cdot 13)(48 \cdot 44 \cdot 40)}{1 \cdot 2 \cdot 3} = 1,098,240$$

The factor $(6 \cdot 13)$ will give the number of ways of getting a pair. The 48 is the possible number of cards not in the pair. The 44 is the possible number of cards not A and not in the pair. The 40 is the possible number of cards not A or B, and not in the pair. For those three cards A, B, and C, we have to eliminate their duplication of order, which is why we write the denominator as $1 \cdot 2 \cdot 3$. The probability of getting one pair is:

$$1,098,240 \div 2,598,960 = 0.4226$$

The chance of getting one pair in a poker hand is 0.4226, or a little better than 4 out of 10. The probability of a pair of Kings is found using the following: The total number of possible pairs of Kings is:

$$\frac{(6 \cdot 1)(48 \cdot 44 \cdot 40)}{1 \cdot 2 \cdot 3} = 8,448$$

The probability of getting a pair of Kings is:

$$\frac{8,448}{2,598,960} = 0.0325 = \frac{325}{10,000}$$

The first product 6 · 1 refers to the fact that there are 6 ways of getting two Kings.

KS, KH   KS, KD   KS, KC
KH, KD   KH, KC   KD, KC

The 1 refers to the fact that we are only considering one face card: the King. This probability of having, say, the pair of Kings is about 1 in 30 chances, whereas the probability of getting *any* pair is 2 out of 5!

Probability and statistics, a large branch of mathematics, is useful in many applications besides gambling. It has applications in the fields of business, economics, genetics, insurance, physics, as well as psychology.

Well, dear reader, I hope you have learned some of the mathematical problem-solving methods presented in this book. You have worked through many everyday applications of percents, graphs, and algebraic formulas. You have been shown shortcuts to work some of the percent problems.

Mathematics is known as the "Queen Science." It embodies many important skills we should know and use to better our lives.

I hope that you have become interested in some of the practical, everyday math problems, and that you will continue your math studies.

INDEX